QUANTITATIVE GEOGRAPHY

QUANTITATIVE GEOGRAPHY

Bimal Dhawan

RANDOM

RANDOM PUBLICATIONS

NEW DELHI - 110 002 (INDIA)

Quantitative Geography

ISBN 978-93-51117-40-7

Published in 2015 in India by

RANDOM PUBLICATIONS

4376-A/4B, Gali Murari Lal, Ansari Road
New Delhi-110 002
Phone: +9111-43580356, 23289044
E-mail: randomexports@gmail.com; sales@randompublications.com; info@randompublications.com

Reprinted 2024

Type Setting by: Friends Media, Delhi-110089
Digitally Printed at: Replika Press Pvt. Ltd.

Preface

Science wouldn't be science without the facts, research, and numbers to back up all those theories. Can you imagine Albert Einstein trying to promote his famous theory of relativity without any numbers, research, or the famous formula E=MC2? The hard numbers behind any good research project are called quantitative data. Quantitative data is the language of science. It uses mathematical models, theories, and hypotheses. In the history of geography, the quantitative revolution (QR) was one of the four major turning-points of modern geography – the other three being environmental determinism, regional geography and critical geography). The quantitative revolution occurred during the 1950s and 1960s and marked a rapid change in the method behind geographical research, from regional geography into a spatial science. The main claim for the quantitative revolution is that it led to a shift from a descriptive (idiographic) geography to an empirical law-making (nomothetic) geography. Proponents of quantitative geography tended to present it as bringing science to geography. In fact, the particular contribution of the quantitative revolution was the huge faith placed in multivariate analysis and in particular methods associated with econometrics. It was also very strongly aligned with positive science, and this would prove a major source of epistemological debate.

The overwhelming focus on statistical modelling would, eventually, be the undoing of the quantitative revolution. Many geographers became increasingly concerned that these techniques simply put a highly sophisticated technical gloss on an approach to study that was barren of fundamental theory. Other critics argued that it removed the 'human dimension' from a discipline that always prided itself on studying the human and natural world alike. As the 1970s dawned, the quantitative revolution came under direct challenge. The greater use of computers in geography also led to many new developments in geomatics, such as the creation and application of GIS and remote sensing. These new developments allowed geographers for the first time to assess complex models on a full-scale model and over space and time. The development of geomatics led to geography being reunited, as the complexities of the human and natural environments could be assessed on new computable models. Further advances also led to a greater role of spatial statistics and modelling within geography. Eventually the quantitative revolution had its greatest impacts on the fields of physical, economic and urban geography. The counter-positivist response from human geography was created in a form of behavioural, radical and humanistic geography. The quantitative revolution also changed the structure of geography departments in the USA, with many physical geographers being merged with geology departments or environmental science departments, leaving the geography departments to become solely human-geography oriented. Within the UK, there was a different response to the revolution, with an increase of specialisation within the subject, and ultimately the development of systematic geography with many subfields and branches. The need to reassess quantitative geography has become only too apparent in recent years as the range of techniques and strategies to be covered has grown.

The present book provides a complete spectrum of views on the achievements of quantitative geography and highlight areas where significant advances are expected to be made.

I thank all members of my team who have helped in the preparation of the book. My special thanks go to "Random Publications" who have published the book.

— *Bimal Dhawan*

Contents

Chapter 1

Introduction

In the history of geography, the quantitative revolution (QR) was one of the four major turning-points of modern geography – the other three being environmental determinism, regional geography and critical geography). The quantitative revolution occurred during the 1950s and 1960s and marked a rapid change in the method behind geographical research, from regional geography into a spatial science. The main claim for the quantitative revolution is that it led to a shift from a descriptive (idiographic) geography to an empirical law-making (nomothetic) geography.

Synopsis and Background

Many geography departments in the 1950s had recently separated from geology departments in the flux of postwar (World War II) enrollment. Because geologists of the time looked at geography as soft and unscientific, the feeling of many geographers was to persuade critics that geographers were not second-rate geologists. The changes during the 1950s through 1970s were not the introduction of mathematics into geography, but mathematics as a tool for explicit purposes and for statistical methodology and formal mathematical modelling.

In the early 1950s, there was a growing sense that the existing paradigm for geographical research was not adequate in explaining how physical, economic, social, and political processes are spatially organised, ecologically related, or how outcomes generated by them are evidence for a given time and place. A more abstract, theoretical approach to geographical research has emerged, evolving the analytical method of inquiry.

The analytical method of inquiry led to the development of generalizations that are logically valid about the spatial aspects of a small set of closely defined events embodied in a wide range of natural and cultural settings. Generalizations may take the form of tested hypotheses, models, or theories, and the research is judged on its scientific fit and its validity. Adoption of the analytical approach had helped geography become a more law-giving science, and the conception of the discipline as an idiographic field of study has become less acceptable starting in the 1980s.

The 1950s Crisis in Geography

During the late 1940s and early 1950s, the crisis occurred for several reasons:

- The closing of many geography departments and courses in universities, e.g., the abolition of the geography program at Harvard University (a highly prestigious institution) in 1948.
- Continuing division between human and physical geography – general talk of human geography becoming an autonomous subject.
- Geography was seen (fairly or not) as overly descriptive and *unscientific* – there was, it was claimed, no explanation of *why* processes or phenomena occurred.
- Geography was seen as exclusively educational – there were few if any applications of contemporary geography.
- Continuing question of what geography is – science, art, humanity or social science?
- After World War II, technology became increasingly important in society, and as a result, nomothetic-based sciences gained popularity and prominence.

Debate raged predominantly (although not exclusively) in the U.S., where regional geography was the major philosophical school (European geography had never been uncomfortable with analytical methods).

All of these events presented a great threat to geography's position as an academic subject, and thus geographers began seeking new methods to counter critique. Under the (somewhat misleading) banner of the scientific method, the *quantitative revolution* began.

The Revolution

The Quantitative Revolution began in the universities of Europe with the support of geographers and statisticians in both Europe and the United States. First emerging in the late 1950s and early 1960s, the Quantitative Revolution responded to the rising regional geography paradigm. Under the loosely defined banner of bringing 'scientific thinking' to geography, the quantitative revolution led to an increased use of computerized statistical techniques, in particular multivariate analysis, in geographical research. The newly adopted methods reflected an array of mathematical techniques that improved precision.

Some of the techniques that epitomize the quantitative revolution include:

- Descriptive statistics;
- Inferential statistics;
- Basic mathematical equations and models, such as gravity model of social physics, or the Coulomb equation;
- Stochastic models using concepts of probability, such as spatial diffusion processes;
- Deterministic models, e.g. Von Thünen's and Weber's location models.

The common factor, linking the above techniques, was a preference for numbers over words, plus a belief that numerical work had a superior scientific pedigree.

Proponents of quantitative geography tended to present it as bringing science to geography. In fact, the particular contribution of the quantitative revolution was the huge faith placed in multivariate analysis and in particular methods associated with econometrics. It was also very strongly aligned with positive science, and this would prove a major source of epistemological debate.

The overwhelming focus on statistical modelling would, eventually, be the undoing of the quantitative revolution. Many geographers became increasingly concerned that these techniques simply put a highly sophisticated technical gloss on an approach to study that was barren of fundamental theory. Other critics argued that it removed the 'human dimension' from a discipline that always prided itself on studying the human and natural world alike. As the 1970s dawned, the quantitative revolution came under direct challenge.

Post-revolution Geography

The greatest impact of the quantitative revolution was not the revolution itself but the effects that came afterwards in a form of the spread of positivist (post-positivist) thinking and counter-positivist responses.

The rising interest in the study of distance as a critical factor in understanding the spatial arrangement of phenomena during the revolution led to formulation of the first law of geography by Waldo Tobler. The development of spatial analysis in geography led to more applications in planning process and the further development of theoretical geography offered to geographical research a necessary theoretical background.

The greater use of computers in geography also led to many new developments in geomatics, such as the creation and application of GIS and remote sensing. These new developments allowed geographers for the first time to assess complex models on a full-scale model and over space and time. The development of geomatics led to geography being reunited, as the complexities of the human and natural environments could be assessed on new computable models. Further advances also led to a greater role of spatial statistics and modelling within geography. Eventually the quantitative revolution had its greatest impacts on the fields of physical, economic and urban geography.

The counter-positivist response from human geography was created in a form of behavioural, radical and humanistic geography.

The quantitative revolution also changed the structure of geography departments in the USA, with many physical geographers being merged with geology departments or environmental science departments, leaving the geography departments to become solely human-geography oriented. Within the UK, there was a different response to the revolution, with an increase of specialisation within the subject, and ultimately the development of systematic geography with many subfields and branches.

Geostatistics

Geostatistics is a branch of statistics focusing on spatial or spatiotemporal datasets. Developed originally to predict probability distributions of ore grades for mining operations, it is currently applied in diverse disciplines including petroleum geology, hydrogeology, hydrology, meteorology, oceanography, geochemistry, geometallurgy, geography, forestry, environmental control, landscape ecology, soil science, and agriculture (esp. in precision farming). Geostatistics is applied in varied branches of geography, particularly those involving the spread of diseases (epidemiology), the practice of commerce and military planning (logistics), and the development of efficient spatial networks. Geostatistical algorithms are incorporated in many places, including geographic information systems (GIS) and the R statistical environment.

Background

Geostatistics is intimately related to interpolation methods, but extends far beyond simple interpolation problems. Geostatistical techniques rely on statistical models that are based on random function (or random variable) theory to model the uncertainty associated with spatial estimation and simulation.

A number of simpler interpolation methods/algorithms, such as inverse distance weighting, bilinear interpolation and nearest-neighbour interpolation, were already well known before geostatistics. Geostatistics goes beyond the interpolation problem by considering the studied phenomenon at unknown locations as a set of correlated random variables.

Let $Z(\mathrm{x})$ be the value of the variable of interest at a certain location x. This value is unknown (e.g. temperature, rainfall, piezometric level, geological facies, etc.). Although there exists a value at location x that could be measured, geostatistics considers this value as random since it was not measured, or has not been measured yet. However, the randomness of $Z(\mathrm{x})$ is not complete, but defined by a cumulative distribution function (CDF) that depends on certain information that is known about the value $Z(\mathrm{x})$:

$$F(z,\mathbf{x}) = \mathrm{Prob}\{Z(\mathbf{x})\leqslant z \mid \mathrm{information}\}.$$

Typically, if the value of Z is known at locations close to x (or in the neighbourhood of x) one can constrain the CDF of $Z(\mathrm{x})$ by this neighbourhood: if a high spatial continuity is assumed, $Z(\mathrm{x})$ can only have values similar to the ones found in the neighbourhood. Conversely, in the absence of spatial continuity $Z(\mathrm{x})$ can take any value. The spatial continuity of the random variables is described by a model of spatial continuity that can be either a parametric function in the case of variogram-based geostatistics, or have a non-parametric form when using other methods such as multiple-point simulation or pseudo-genetic techniques.

By applying a single spatial model on an entire domain, one makes the assumption that Z is a stationary process. It means that the same statistical properties are applicable on the entire domain. Several geostatistical methods provide ways of relaxing this stationarity assumption.

In this framework, one can distinguish two modelling goals:

1. Estimating the value for $Z(\mathrm{x})$, typically by the expectation, the median or the mode of the CDF $f(z,\mathrm{x})$. This is usually denoted as an estimation problem.
2. Sampling from the entire probability density function $f(z,\mathrm{x})$ by actually considering each possible outcome of it at each location. This is generally done by creating several alternative maps of Z, called realisations. Consider a domain discretized in N grid nodes (or pixels). Each realisation is a sample of the complete N-dimensional joint distribution function

$$F(\mathbf{z},\mathbf{x}) = \text{Prob}\{Z(\mathbf{x}_1)\leqslant z_1, Z(\mathbf{x}_2)\leqslant z_2,\ldots, Z(\mathbf{x}_N)\leqslant z_N\}.$$

In this approach, the presence of multiple solutions to the interpolation problem is acknowledged. Each realisation is considered as a possible scenario of what the real variable could be. All associated workflows are then considering ensemble of realisations, and consequently ensemble of predictions that allow for probabilistic forecasting. Therefore, geostatistics is often used to generate or update spatial models when solving inverse problems.

A number of methods exist for both geostatistical estimation and multiple realisations approaches. Several reference books provide a comprehensive overview of the discipline.

Geostatistical Conditional Simulation

Geostatistical simulation is well accepted in the petroleum industry as a method for characterizing heterogeneous reservoirs. It often is preferable to traditional interpolation approaches, in part because it captures the heterogeneous character observed in many petroleum reservoirs and provides more accurate hydrocarbon reserve estimates. Geostatistical simulation methods preserve the variance observed in the data, instead of just the mean value, as in interpolation. Their stochastic approach allows calculation of many equally probable solutions (realisations), which can be post-processed to quantify and assess uncertainty.

Use of Stochastic Approach

Many practitioners are suspicious of stochastic methods—and even reject them outright—because natural processes that form reservoirs are not random. But geostatistical stochastic simulation is not a coin-toss experiment. Furthermore, while it is true that reservoirs are not products of random processes, it also is true that they have attributes that cause them to behave as if they were random. For example, physical and chemical processes modify reservoir characteristics from their original state, confounding our ability to make predictions even when we understand the processes. Such changes cause behaviour that can be captured by stochastic principles.

Kriging is a deterministic method whose function has a unique solution and does not attempt to represent the actual variability of the studied attribute. The smoothing property of any interpolation algorithm replaces local detail with a good average value; however, the geologist and reservoir engineer are more interested in finer-scaled details of reservoir heterogeneity than in a map of local estimates of the mean value. Like the traditional deterministic approach, stochastic methods preserve hard data where known and soft data where informative. Unlike the deterministic approach, though, it provides geoscientists and reservoir engineers with many realisations. The kriged solution is the average of numerous realisations, and the variability in the different outcomes is a measure of uncertainty at any location. Thus, the standard deviation of all values simulated at each grid node is the quantification of uncertainty.

What Do We Want from a Simulation?

Which simulation method we choose depends on what we want from a stochastic-modelling effort and—to great extent—the types of available data. Not all conditional simulation studies need a "Cadillac" method. For many, a "Volkswagen" serves the purpose well. Among the reasons for performing stochastic simulation, four important ones are:

1. To capture heterogeneity
2. To simulate facies or petrophysical properties, or both
3. To honour and integrate multiple data types
4. To quantify and assess uncertainty

Principles of Stochastic Modelling

In general, conditional simulation requires that the basic input parameters—the spatial model (variograms) and the distribution of sample values (cumulative distribution function, or cdf)—remain constant within a given geologic interval and/or facies, from realisation to realisation. Typically, the structural and stratigraphic model (major structural surfaces and the discretized layers between them) remains fixed. Because each realisation begins with a different, random seed number, each has a unique "random walk," or navigational path through the 3D volume. The random walk provides the simulation algorithm with the order of cells to be simulated, and is different from realisation to realisation; therefore, the results are different at unsampled locations, producing local changes in the distribution of facies and petrophysical properties in the interwell space. Note that selection of the same random seed always will reproduce the same random walk. This characteristic is for computational convenience. In practice, multiple realisations are performed at or close to the geologic scale, and not necessarily at the flow-simulation scale.

There are two basic categories of conditional simulation methods:

- Pixel-based methods operate on one pixel at a time. They can be used with either continuous or categorical data.
- Object-based methods operate on groups of pixels that are connected and arranged to represent genetic shapes of geologic features. They are used only with categorical data.

Pixel Based Model

A pixel-based model assumes that the variable to be simulated is a realisation of a continuous (Gaussian) random function. Using the spatial model, search ellipse, and control data, a pixel-based method simulates values grid node by grid node. Some of the most popular pixel-based algorithms are: turning bands, sequential Gaussian, sequential indicator, truncated Gaussian, and simulated annealing. Each method can produce a range of realisations that capture the uncertainty of an regionalized variable (RV), and so the method choice here will be based on the goals and on data types and their availability. The pixel-based method works best in the presence of facies associations that vary smoothly across the reservoir, as often is the case in deltaic or shallow marine reservoirs. No assumption is made about the shape of the sedimentary bodies. This method is preferred when the net-to-gross ratio is high.

Object Based (Boolean) Model

The algorithms for a Boolean model generate spatial distributions of sedimentary bodies (channels, crevasse splays, reefs, etc.) whose parameters (orientation, sinuosity, length, width, etc.) can be inferred from the assumed depositional model, seismic data, outcrops, and even well-test interpretations. The object-based method simulates many grid nodes at one time, superimposing geometries (e.g., sheets, discs, or sinusoids) onto a background that typically is a shaly lithofacies. The method used for object modelling is referred to as the marked-point process. This method generally works best with a low net-to-gross ratio and widely spaced wells.

Selecting the Right Method

It is difficult to say *a priori* which type of method, pixel- or object-based, is best. Although we have observed that it is common for practitioners to have a personal bias toward one method or the other, the basis for such preference often is not well founded. For example, method selection often is based on familiarity with the procedure or what is available in a given software package. Additionally, we have observed that geologists tend to prefer object-based methods because they often produce realisations that appear "crisp" and realistic (e.g., deltas look like deltas and channels look like channels). Engineers tend toward pixel-based methods because they require less descriptive input and often are computationally faster than object-based methods. In fact, both methods are computationally sound and offer unique characteristics.

From a practical point of view, the methods can be combined to achieve an effective model. For example, the practitioner could model a transition sequence from offshore to onshore with a pixel-based method, and then superimpose a channel system from a delta using an object-based method. To help in selecting the appropriate methods, we can offer the following for consideration:

Pixel-based methods are more forgiving in that they require fewer assumptions about the data. As such, the error variance generated from a set of realisations generally will be higher than with object-based modelling. We surmise that pixel-based models create a larger space of uncertainty and therefore are more likely to "capture" the correct solution, even if the initial conceptual geologic model is incorrect.

Object-based models work best when the data density and net-to-gross ratio are low. A set of object realisations will generate a lower error variance than that from a pixel-based model, and thus can be said to have a smaller space of uncertainty. When the conceptual geologic model is strongly supported by the data and is well understood, the method is highly successful; however, because more assumptions about the data are required, the resultant realisations are less forgiving (i.e., if the original conceptual model is wrong, there is little chance of it successfully capturing the correct solution).

Stochastic Simulation Methods

There are several theoretically sound and practically tested conditional-simulation approaches, and choosing one can be bewildering and daunting for a novice to stochastic methods. Parametric-simulation techniques assume that the data have a Gaussian distribution, so that a transform of the data typically is prerequisite. Note that indicator data do not undergo such a transform. Furthermore, data transformations are not required in the object-based method, where only indicator data are required. The steps of parametric simulation are:

- Perform a normal-score data transform from the z -space to the y -space.
- Compute and model the variogram (covariance) on the y- normal scores.
- Perform multiple simulations of the y- normal scores on a grid or within a volume.
- Back-transform the simulated y- normal scores to the simulated z -values.
- Post-process the multiple simulations to assess uncertainty.

Turning Bands Simulation

In turning bands simulation (one of the earliest simulation methods), first the data are kriged and then unconditional simulations are created using a set of randomly distributed bands, or lines. The general procedure is as follows:

1. Raw data values are kriged to a regular grid.
2. Numerous random lines (bands) with various azimuths are generated around a centroid located at the grid or volume centre. The modeller controls the number of lines.
3. Unconditional simulations of normal-score transformed data are performed along each line using the transformed-data histogram and variogram.
4. Values along the lines are linearly interpolated to grid nodes—the more lines, the less interpolation.
5. Unconditional interpolated values are back interpolated to well locations.
6. Unconditional interpolated values and well locations are kriged.
7. The grid of unconditional interpolated values is subtracted from the grid of unconditional results in step 5. This creates a residual map with a value of zero at the well locations.
8. The residuals are back interpolated from the y-space to the z-space.
9. The back-interpolated residuals from step 8 are added to the original kriged map from step 1.

10. The result is a grid or volume of data values that reproduce both the mean of the raw data and the variance.

Sequential Simulation

Three sequential-simulation procedures use the same basic algorithm for different data types:

- Sequential Gaussian simulation (SGS) simulates continuous variables, such as petrophysical properties
- Sequential indicator simulation (SIS) simulates discrete variables, using SGS methodology to create a grid of zeros and ones
- Bayesian indicator simulation (a newer form of SIS) allows direct integration of seismic attributes, and uses a combination of classification and indicator methods.

As described in the literature, the general sequential simulation process is :

1. Perform a normal-score transformation of the raw data.
2. Randomly select a node that is not yet simulated in the grid.
3. Estimate the local conditional probability distribution function (lcpd) for the residuals at the selected node. The residuals can be calculated by subtracting the grid of an unconditional simulation from a kriged grid of the unconditional values sampled at the geographic coordinates of the wells.
4. Create a newly simulated value by adding together the randomly drawn residual value and the mean of the transformed data.
5. Include the newly simulated value in the set of conditioning data, within a specified radius of the new target location. This ensures that closely spaced values have the correct short-scale correlation.
6. Repeat until all grid nodes have a simulated value.

As with turning-bands simulation, each time a new random walk is defined, a new and different result will occur. In this case, though, the lcpd is updated continually by the previously simulated values.

Truncated Gaussian Simulation

Developed by Institut Français du Pétrole and the Centré de Géostatistiques, the truncated Gaussian algorithm simulates lithofacies directly by using a set of cutoffs that partition Gaussian field. The cutoffs commonly are generated from facies proportions calculated from well data. One simple method for doing this is by calculating the vertical proportion curve. A vertical proportion curve is a stacked bar diagram that represents the percentage of all facies in all wells in the study area. The proportion of each facies is calculated layer by layer, where a layer is a subdivision of the reservoir unit being modelled. Truncated Gaussian simulation involves first generating a continuous variable, and then applying cutoffs (termed the Gaussian thresholds) during the simulation. This method works exceptionally well with transitional facies, such as those from foreshore to upper shoreface to lower shoreface to offshore.

Simulated Annealing Simulation

The simulated-annealing simulation method is borrowed from metallurgy. In metallurgy, when fusing two pieces of metal, the attachment zone is heated to a temperature at which the molecular structure can be rearranged. As the metal cools again, the molecular structure changes and a bond forms where the two pieces of metal are joined. In transferring this idea to stochastic modelling, one produces an initial realisation, introduces some particular conditions (new pieces of "metal" to be fused), then "heats" and "cools" it to rearrange the pixels (or objects) to match (band) the particular conditions introduced.

The simulated-annealing simulation method constructs the reservoir model through an iterative, trial-and-error process, and does not use an explicit random-function model. It can be used as the basis for both pixel- and object-based simulation and in either case the simulated image is formulated as an optimization process. For example, our desired result might be an image of a sand/shale model with a 70% net-to-gross ratio, an average shale length of 60 m, and average shale thicknesses of 10 m. The starting image has pixels (objects) arranged randomly, with the correct global proportion of sand and shale, but with an incorrect net-to-gross relationship that stems from the completely random assignment of the sand and shale. In addition, the average shale lengths and widths are not correct. During the computation, the annealing algorithm modifies the initial image by swapping information from node to node, and determining whether or not an individual swap improves the realisation. This method allows some swap "improvements" to be rejected to prevent the occurrence of a "local minimum," a well-known problem with annealing techniques. The swapping process continues until a final image is produced that matches the statistics of the input data.

The simulated-annealing process produces excellent results, but can be inefficient because millions of perturbations may be required to arrive at the desired image. Nevertheless, the availability of faster computers with more memory are making simulated-annealing simulation methods more attractive. They are particularly desirable for integrating dynamic data, such as production histories and well tests, to ensure history matching from any realisation.

Boolean simulation methods create reservoir models based on objects that have a genetic significance, rather than building up models one elementary node or pixel at a time. To use Boolean methods, select a basic shape for each depositional facies that describes its geometry. For example, you might model channels that look sinuous in map view and half-elliptical in cross section, or deltas that look like triangular wedges in map view. The modeler must specify the proportions of shapes in the final model and choose parameters that describe the shapes. Some algorithms have rules describing how geologic bodies are positioned relative to each other. Once the shape distribution parameters and position rules are chosen, the six-step simulation is performed:

1. Fill the reservoir model background with a lithofacies, such as shale.
2. Randomly select a starting point in the model.
3. Randomly select one lithofacies shape and draw it according to the shape, size, and orientation rules.
4. Check to see whether the shape conflicts with any control data or with previously simulated shapes. If it does, reject it and go back to step 3.
5. Check to see whether the global lithofacies proportions are correct. If they are not, return to step 2.
6. Use pixel-based methods to simulate petrophysical properties within the geologic bodies.

If control data must be honoured, objects are simulated at control points first, before simulating the interwell region. Be sure that there are no conflicts with known stratigraphic and lithologic sequences at the well locations.

Boolean techniques currently are of interest to the petroleum industry, and geologists find object modelling particularly satisfactory because the shapes created are based on geometries and facies relationships that have been measured, and because the images look geologically realistic. Criticisms of Boolean modelling are that it requires a large number of input parameters and that prior knowledge is needed to select the parameter values. Furthermore, in the past,

Boolean-type algorithms could not always honour all the control data because the algorithms are not strict simulators of shape, and often require changes to the facies proportions of object sizes to complete a given realisation; however, new technology has greatly alleviated this problem. A number of research/academic institutions and vendors continue to research better ways to implement these algorithms.

Uncertainty Quantification and Assessment

As discussed earlier, stochastic models allow quantification of uncertainty related to the geologic description. An infinite number of possible realisations are obtained simply by modifying the seed number. Comparing a sufficiently large number of these realisations then will provide a measurement of the uncertainty associated with the assumed geologic model.

Since the late 1990s, real modelling experiences have generated questions concerning the workflow, aimed at capturing and developing a working definition for uncertainty. Currently, for example, the accepted implementation of stochastic modelling involves four general steps:

- Produce multiple realisations of the fine-scaled model by changing the seed number
- Rank the results on the basis of some criteria
- Upscale the P10, P50, and P90 results
- Flow-simulate the above three solutions to capture the range of the uncertainty

One common criticism of the above workflow is that the actual space of uncertainty is much larger than that explored by the variability of the random function. This concept often is overlooked; we tend to identify uncertainty on the basis of stochastic simulations that fix all input parameters, and simply to change the seed value from simulation to simulation. There are at least five sources of uncertainty in a typical reservoir model:

1. Uncertainty in data quality and interpretation—Basic data-measurement errors generally are ignored and the data are treated as error-free when modelling the reservoir. The same holds true for interpreted data, such as potential errors in picking seismic time horizons.
2. Uncertainty in the structural model—The structural model virtually always is created using a deterministic approach (e.g., when seismic time horizons are converted to depth using an

uncertain velocity model, then treated as fixed surfaces in the geologic model.) The structural model is one of the largest sources of uncertainty, and greatly affects volumetric calculations.

3. Uncertainty in the stratigraphic model—Uncertainty in the stratigraphic model is related to the reliability of sequence determination and correlation through the wells.
4. Uncertainty in the stochastic model choice and its parameters—If the same geologic sequence were modelled using different algorithms, each stochastic simulation method would yield different results and explore a different part of the space of uncertainty; however, the space sampled by the different algorithms would overlap significantly.
5. Uncertainty from multiple realisations—The uncertainty reported in most stochastic-modelling studies usually is from multiple realisations. Compared to the other sources for error, it often is small (but not always).

The above sources of uncertainty are listed in decreasing order of significance. It should be somewhat intuitive that change in the data, structural model, or sequence stratigraphic model is likely to have a greater impact on the reservoir model than will changing a single parameter of a given random function; however, it is important to understand that when building a geologic model, all levels of uncertainty must be accounted for to achieve an accurate assessment of the space of uncertainty. The next section further defines the order of uncertainty and the relationship between scenarios and realisations.

Realisations, Scenarios, and the Space of Uncertainty

To account for the different sources of uncertainty, we can classify uncertainty into three orders on the basis of the degree of impact on the reservoir model.

- First-order uncertainty stems from major changes in modelling assumptions, such as changing the data or the structural model, testing of different depositional models, or changing the petrophysical model.
- Second-order uncertainty is caused by small changes to the parameter of the random function (e.g., the sill, range, or model of the variogram), to the cumulative distribution functions (cdfs), or to the stochastic-model choice.
- Third-order uncertainty results from changes only in the interwell space that are caused by the algorithm selected with

its parameterization, and by changing the seed number from realisation to realisation. (parameters that control first- and second-order uncertainties remain fixed.) We refer to the larger changes in uncertainty that are represented by the first and second order as scenarios, whereas we call the third order changes realisations.

Quite often, scenario modelling plays a key role early in the field development phase, when data are sparse and operators hold fundamental differences of opinion about the appropriate conceptual depositional model to use. For example, they might differ on whether the system is wave-dominated fluvial deltaic or tidal-dominated. Although both are deltaic, the overprint of the oceanic processes changes the strike of dominant sand bodies, making it either parallel to the coast (wave-dominated) or perpendicular to the coast (tidal-dominated). Note that each scenario model will require its own set of realisations. Thus, measuring the total space of uncertainty will require that multiple scenarios and their respective realisations be constructed and their ranges of uncertainty be pooled together. This may cause a great deal of computational effort, but it often is a mistake to assume that by modelling the scenarios without their respective realisations, the critical spectrum of uncertainty will be captured.

Static Displays of Uncertainty

The most common way to visualize uncertainty is as a static view, through summary statistics maps prepared from a suite of realisations. Several types of displays are conventionally are used for this purpose. These include:

- Maps of mean and median
- Maps of spread
- Uncertainty/probability/risk maps
- Isoprobability maps
- Display of multiple realisations

Maps of Mean and Median

Mean and median maps are based on the average and median of a given number of conditional simulations. At each cell, the program computes the average or median value for the values at that location from all simulations. When the number of simulations is large, the map converges to the kriged solution. Mean and median maps are identical when they are based on the statistics of an infinite number of simulations.

Maps of Spread

Spread is most commonly displayed as a map of the standard deviation (error) at each grid cell, computed from all input maps.

Maps of Uncertainty, Probability and Risk

The probability of meeting or exceeding a user-specified threshold at each grid cell is displayed using maps of uncertainty, probability, and risk, in which grid-cell values range from 0 to 100%. During post-processing, the modeler fixes the threshold, and the program computes the probability of meeting or exceeding it.

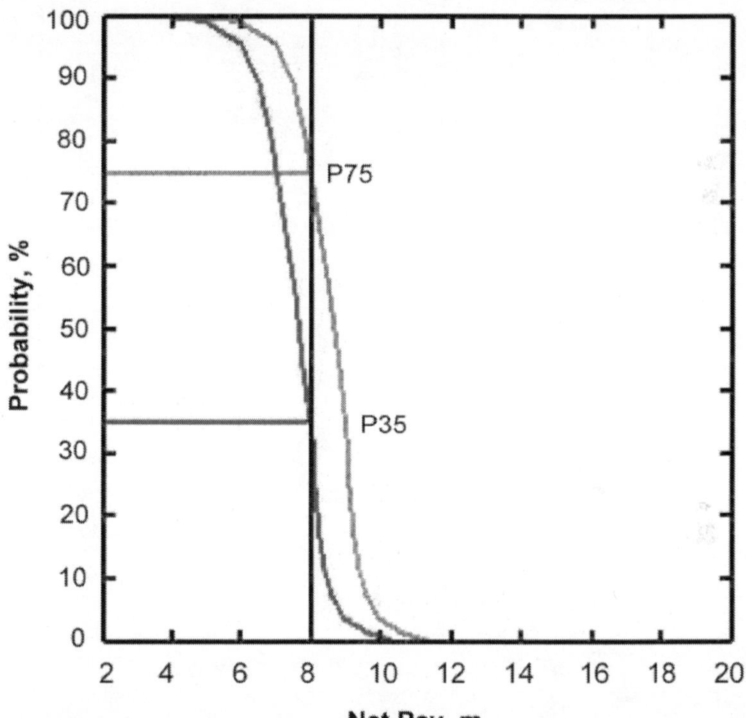

Figure: *—Plot showing the probability of finding at least 8 m net pay. The location represented by the lefthand curve shows only a 35% chance (P35) of encountering 8 m or more of net pay, but the location for the right-hand curve has a 75% chance (P75) of meeting that economic criterion.*

Isoprobability Maps

Rather than holding the threshold constant, sometimes it is preferable to freeze the probability value and create maps that correspond to local quantiles or percentiles. Such maps are known as isoprobability maps, and their grid-cell values relate to the attribute, rather than to probability. An isoprobability map created at the tenth percentile shows that net pay has a 10% chance of being thick, because the modelled variable is metres; this represents a pessimistic view of the hydrocarbon potential. Conversely, a ninetieth percentile is an optimistic picture of the hydrocarbon potential, showing that there is only a 10% chance that the net pay is very thin.

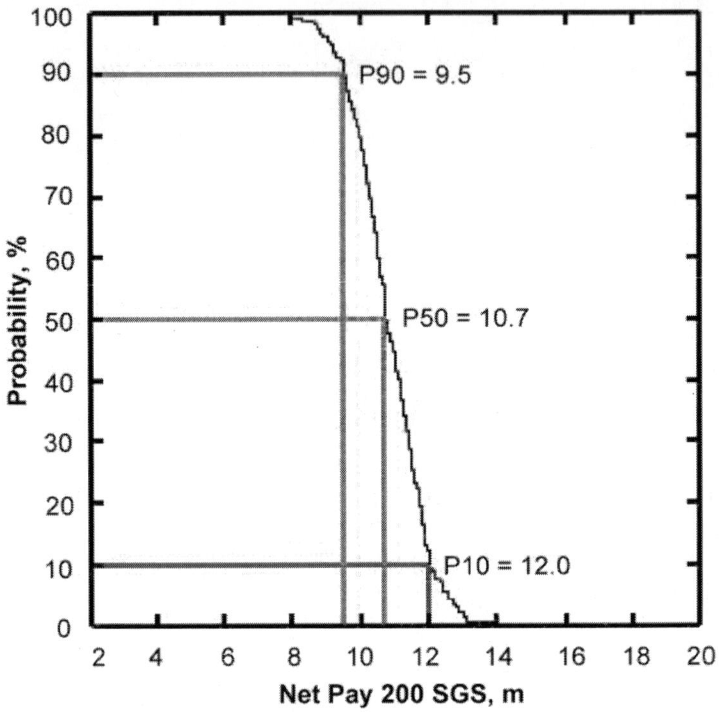

Figure: *—Probability plot for net pay. Isoprobability maps are constructed from such plots by freezing the probability values and determining the value of an attribute. The probability distribution shows the range of values simulated at a single grid node. Values for P90, P50, and P10 are shown.*

Multiple Realisations

Another common format for illustrating uncertainty is simply to display several possible realisations that as a group represent the broad spectrum of outcomes.

Dynamic Displays of Uncertainty

In his discussions on the use of animated (dynamic) displays of the realisations, Srivastava emphasizes that, like a well-produced animated cartoon, a visually effective dynamic display of uncertainty should present the realisations in a gradual and logically successive and informative way (i.e., not simply in random succession). The key to a successful dynamic rendering of uncertainty is finding a way to show all the realisations.

Separate displays of realisations would be a step in the right direction, but workstation screen space is limited, as is the patience of the modeler scrolling through the images. The scrolling process could be automated, presenting the realisations one at a time in rapid succession, but successive realisations might differ considerably in local detail, and create a flickering display that would be more irritating than illuminating.

To make an appealing dynamic display, each realisation must be treated as if it were a frame from a movie clip, with each successive frame showing minimal difference from the last, so that the brain interprets the minor transitions as gradual movement.

For example, were we to create realisations of a structural top and make an animated display, the movie clip might show a perpetually undulating dome-shaped feature. The surface would appear fixed at well locations, and thus stable. Moving away from the well control, the surface would appear to swell and heave; small depressions would grow into large, deep holes and valleys. Small bumps would grow into hills and then into larger hills with ridges, and then would shrink again. In several seconds of animation, the modeler could see several hundred realisations, and the variability (uncertainty) in the suite of realisations would show in the magnitude of changes over time. The modeler instantly would recognise areas that show little movement (more certainty) vs. those that wax and wane (greater uncertainty).

In making dynamic displays, though, one must overcome the fact that most geostatistical simulation algorithms have no way to produce a series of realisations that show minimal differences, which animation requires. Recall that to generate a new simulated result, we select a new seed number for the random number generator and rerun the program. But any change in the seed number, however small, produces unpredictable changes in the appearance of the resulting simulation, which means that consecutive seed numbers, for example, could produce simulations that are quite different in their local variability. There is no way, then, to predict which seed numbers will produce similar-enough realisations to construct consecutive frames for the animation. One way we overcome this problem is with a simulation technique known as probability field simulation (P-field simulation), although this technique has its own advantages and disadvantages.

P-Field Simulation

P-field simulation is a conditional simulation technique developed by Froidevaux and Srivastava. The advantage of P-field simulation is that it is ideally suited to the problem of uncertainty animation. It sets up a matrix of probabilities with dimensions that are identical to the 2D or 3D project grid. The spatial model controls the pattern of the probabilities on the matrix; that is, a value of high probably most likely will be adjacent to another high value, and such values could be arranged along the direction of continuity if the variogram is anisotropic.

To generate a new realisation, one only needs to shift the values in the probability matrix by one row or one column. It is not necessary to generate a new random seed and a subsequent random walk. The result is a small, incremental change from the previous realisation. Interestingly, any conditional simulation method that uses unconditional simulation as an intermediate step can be configured to produce a set of ordered realisations that show a progression of small changes from one to the next and that can be animated.

If given viable techniques, geoscientists and reservoir engineers will be able to design and plan more at the workstation. Consider how much more effective a reservoir-development team could be using such techniques—creating a dynamic display for a well in the thickest part of the reservoir, for example, and discovering from the animation that there exists no predictably economically viable net pay. Such techniques could bring reality to the reservoir-modelling process and ensure that planning does not take place on a single, arbitrary model that cannot duplicate reality.

Markov Chain Geostatistics

Markov chain geostatistics refer to the Markov chain models, simulation algorithms and associated spatial correlation measures (e.g., transiogram) based on the Markov chain random field theory, which extends a single Markov chain into a multi-dimensional field for geostatistical modelling. A Markov chain random field is still a single spatial Markov chain. The spatial Markov chain moves or jumps in a space and decides its state at a location through interactions with its nearest known neighbours in different directions, including its last stay location. Because single-step transition probability matrices are difficult to estimate from sparse sample data and are impractical in representing the complex spatial heterogeneity of states, the transiogram, which is defined as a transition probability function over the distance lag, is proposed as the accompanying spatial measure of Markov chain random fields.

Markov Random Field

In the domain of physics and probability, a Markov random field (often abbreviated as MRF), Markov network or undirected graphical model is a set of random variables having a Markov property described by an undirected graph. A Markov random field is similar to a Bayesian network in its representation of dependencies; the differences being that Bayesian networks are directed and acyclic, whereas Markov networks are undirected and may be cyclic. Thus, a Markov network can represent certain dependencies that a Bayesian network cannot (such as cyclic dependencies); on the other hand, it can't represent certain dependencies that a Bayesian network can (such as induced dependencies).

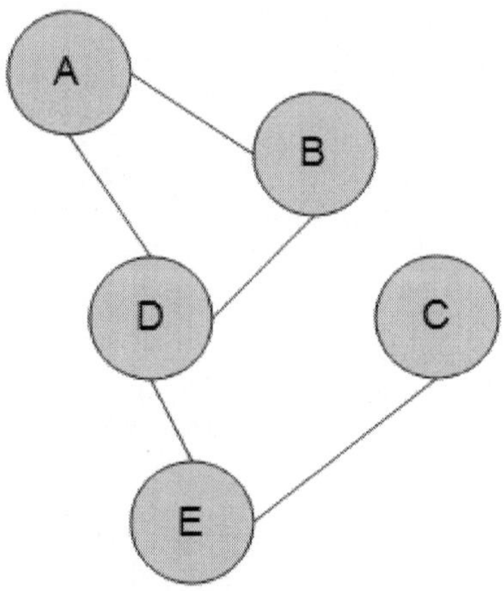

Figure: *An example of a Markov random field. Each edge represents dependency. In this example: A depends on B and D. B depends on A and D. D depends on A, B, and E. E depends on D and C. C depends on E.*

When the probability distribution is strictly positive, it is also referred to as a Gibbs random field, because, according to the Hammersley–Clifford theorem, it can then be represented by a Gibbs measure. The prototypical Markov random field is the Ising model; indeed, the Markov random field was introduced as the general setting for the Ising model. In the domain of artificial intelligence, a Markov random field is used to model various low- to mid-level tasks in image processing and computer vision. For example, MRFs are used for image restoration, image completion, segmentation, image registration, texture synthesis, super-resolution, stereo matching and information retrieval.

Definition

Given an undirected graph $G = (V, E)$, a set of random variables $X = (X_v)_{v \in V}$ indexed by V form a Markov random field with respect to G if they satisfy the local Markov properties:

Pairwise Markov property: Any two non-adjacent variables are conditionally independent given all other variables:

$$X_u \perp\!\!\!\perp X_v \mid X_{V \setminus \{u,v\}} \quad \text{if } \{u,v\} \notin E$$

Local Markov property: A variable is conditionally independent of all other variables given its neighbours:

$$X_v \perp\!\!\!\perp X_{V \setminus \mathrm{cl}(v)} \mid X_{\mathrm{ne}(v)}$$

where ne(v) is the set of neighbours of v, and cl(v) = $\{v\} \cup$ ne(v) is the closed neighbourhood of v.

Global Markov property: Any two subsets of variables are conditionally independent given a separating subset:

$$X_A \perp\!\!\!\perp X_B \mid X_S$$

where every path from a node in A to a node in B passes through S.

The above three Markov properties are not equivalent to each other at all. In fact, the Local Markov property is stronger than the Pairwise one, while weaker than the Global one.

Clique Factorization

As the Markov properties of an arbitrary probability distribution can be difficult to establish, a commonly used class of Markov random fields are those that can be factorized according to the cliques of the graph.

Given a set of random variables $X = (X_v)_{v \in V}$, let $P(X = x)$ be the probability of a particular field configuration x in X. That is, $P(X = x)$ is the probability of finding that the random variables X take on the particular value x. Because X is a set, the probability of x should be understood to be taken with respect to a *joint distribution* of the X_v.

If this joint density can be factorized over the cliques of G:

$$P(X = x) = \prod_{C \in \mathrm{cl}(G)} \phi_C(x_C)$$

then X forms a Markov random field with respect to G. Here, cl(G) is the set of cliques of G. The definition is equivalent if only maximal cliques are used. The functions φ_C are sometimes referred to as *factor potentials* or *clique potentials*. Note, however, conflicting terminology is in use: the word *potential* is often applied to the logarithm of φ_C. This is because, in statistical mechanics, $\log(\varphi_C)$ has a direct interpretation as the potential energy of a configuration x_C.

Although some MRFs do not factorize (a simple example can be constructed on a cycle of 4 nodes), in certain cases they can be shown to be equivalent conditions:

- if the density is positive (by the Hammersley–Clifford theorem),
- if the graph is chordal (by equivalence to a Bayesian network).

When such a factorization does exist, it is possible to construct a factor graph for the network.

Logistic Model

Any Markov random field (with a strictly positive density) can be written as log-linear model with feature functions f_k such that the full-joint distribution can be written as

$$P(X = x) = \frac{1}{Z}\exp\left(\sum_k w_k^\top f_k(x_{\{k\}})\right)$$

where the notation

$$w_k^\top f_k(x_{\{k\}}) = \sum_{i=1}^{N_k} w_{k,i} \cdot f_{k,i}(x_{\{k\}})$$

is simply a dot product over field configurations, and Z is the partition function:

$$Z = \sum_{x \in \mathcal{X}} \exp\left(\sum_k w_k^\top f_k(x_{\{k\}})\right).$$

Here, $\mathcal{X}$ denotes the set of all possible assignments of values to all the network's random variables. Usually, the feature functions $f_{k,i}$ are defined such that they are indicators of the clique's configuration, *i.e.* $f_{k,i}(x_{\{k\}}) = 1$ if $x_{\{k\}}$ corresponds to the *i*-th possible configuration of the *k*-th clique and 0 otherwise. This model is equivalent to the clique factorization model given above, if $N_k = |\operatorname{dom}(C_k)|$ is the cardinality of the clique, and the weight of a feature $f_{k,i}$ corresponds to the logarithm of the corresponding clique factor, *i.e.*. $w_{k,i} = \log \phi(c_{k,i})$, where $c_{k,i}$ is the *i*-th possible configuration of the *k*-th clique, *i.e.* the *i*-th value in the domain of the clique.

The probability P is often called the Gibbs measure. This expression of a Markov field as a logistic model is only possible if all clique factors are non-zero, *i.e.* if none of the elements of are assigned a probability of 0. This allows techniques from matrix algebra to be applied, *e.g.* that the trace of a matrix is log of the determinant, with the matrix representation of a graph arising from the graph's incidence matrix.

The importance of the partition function Z is that many concepts from statistical mechanics, such as entropy, directly generalize to the case of Markov networks, and an *intuitive* understanding can thereby be gained. In addition, the partition function allows variational methods to be applied to the solution of the problem: one can attach a driving force to one or more of the random variables, and explore the reaction of the network in response to this perturbation. Thus, for example, one may add a driving term J_v, for each vertex v of the graph, to the partition function to get:

$$Z[J] = \sum_{x \in \mathcal{X}} \exp\left(\sum_k w_k^\top f_k(x_{\{k\}}) + \sum_v J_v x_v \right)$$

Formally differentiating with respect to J_v gives the expectation value of the random variable X_v associated with the vertex v:

$$E[X_v] = \frac{1}{Z} \frac{\partial Z[J]}{\partial J_v}\bigg|_{J_v=0}.$$

Correlation functions are computed likewise; the two-point correlation is:

$$C[X_u, X_v] = \frac{1}{Z} \frac{\partial^2 Z[J]}{\partial J_u \partial J_v}\bigg|_{J_u=0, J_v=0}.$$

Log-linear models are especially convenient for their interpretation. A log-linear model can provide a much more compact representation for many distributions, especially when variables have large domains. They are convenient too because their negative log likelihoods are convex. Unfortunately, though the likelihood of a logistic Markov network is convex, evaluating the likelihood or gradient of the likelihood of a model requires inference in the model, which is in general computationally infeasible.

Examples

Gaussian Markov Random Field: A multivariate normal distribution forms a Markov random field with respect to a graph

$G = (V, E)$ if the missing edges correspond to zeros on the precision matrix (the inverse covariance matrix):

$$X = (X_v)_{v \in V} \sim \mathcal{N}(\boldsymbol{\mu}, \Sigma)$$

such that

$$(\Sigma^{-1})_{uv} = 0 \quad \text{if} \quad \{u, v\} \notin E.$$

Inference

As in a Bayesian network, one may calculate the conditional distribution of a set of nodes $V' = \{v_1, \ldots, v_i\}$ given values to another set of nodes $W' = \{w_1, \ldots, w_j\}$ in the Markov random field by summing over all possible assignments to $u \notin V', W'$; this is called exact inference. However, exact inference is a #P-complete problem, and thus computationally intractable in the general case. Approximation techniques such as Markov chain Monte Carlo and loopy belief propagation are often more feasible in practice. Some particular subclasses of MRFs, such as trees, have polynomial-time inference algorithms; discovering such subclasses is an active research topic. There are also subclasses of MRFs that permit efficient MAP, or most likely assignment, inference; examples of these include associative networks. Another interesting sub-class is the one of decomposable models (when the graph is chordal): having a closed-form for the MLE, it is possible to discover a consistent structure for hundreds of variables.

Conditional Random Fields

One notable variant of a Markov random field is a conditional random field, in which each random variable may also be conditioned upon a set of global observations o. In this model, each function ϕ_k is a mapping from all assignments to both the clique *k* and the observations to the nonnegative real numbers. This form of the Markov network may be more appropriate for producing discriminative classifiers, which do not model the distribution over the observations.

Support Vector Machine

In machine learning, support vector machines (SVMs, also support vector networks) are supervised learning models with associated learning algorithms that analyze data and recognise patterns, used for classification and regression analysis. Given a set of training examples, each marked as belonging to one of two categories, an SVM training algorithm builds a model that assigns new examples into one category or the other, making it a non-probabilistic binary linear classifier. An SVM model is a representation of the examples as points in space, mapped so that the examples of the separate categories are divided by a clear gap that is as wide as possible. New examples are then mapped into that same space and predicted to belong to a category based on which side of the gap they fall on.

In addition to performing linear classification, SVMs can efficiently perform a non-linear classification using what is called the kernel trick, implicitly mapping their inputs into high-dimensional feature spaces.

Data Mining Algorithms

Support Vector Machines (SVMs) are supervised learning methods used for classification and regression tasks that originated from statistical learning theory. As a classification method, SVM is a global classification model that generates non-overlapping partitions and usually employs all attributes. The entity space is partitioned in a single pass, so that flat and linear partitions are generated. SVMs are based on maximum margin linear discriminants, and are similar to probabilistic approaches, but do not consider the dependencies among attributes.

Traditional Neural Network approaches have suffered difficulties with generalization, producing models which overfit the data as a consequence of the optimization algorithms used for parameter selection and the statistical measures used to select the best model. SVMs have been gaining popularity due to many attractive features and promising empirical performance. They are based on the Structural Risk Minimization (SRM) principle have shown to be superior to the traditional principle of Empirical Risk Minimization (ERM) employed by conventional Neural Networks. ERM minimizes the error on the training data, while SRM minimizes an upper bound on the expected risk. This gives SRM greater generalization ability, which is the goal in statistical learning. According to , SVMs rely on preprocessing the data to represent patterns in a high dimension, typically much higher than the original feature space. Data from two categories can always be separated by a hyperplane when an appropriate nonlinear mapping to a sufficiently high dimension is used.

A classification task usually involves training and test sets which consist of data instances. Each instance in the training set contains one target value (class label) and several attributes (features). The goal of a classifier is to produce a model able to predict target values of data instances in the testing set, for which only the attributes are known. Without loss of generality, the classification problem can be viewed as a two-class problem in which one's objective is to separate the two classes by a function induced from available examples. The goal is to produce a classifier that generalizes well, i.e. that works well on unseen examples. The below picture is an example of a situation in which various linear classifiers can separate the data. However, only one maximizes the distance between itself and the nearest example of each class (i.e. the margin) and for that is called the optimal separating hyperplane. It is intuitively expected that this classifier generalizes better than the other options. The basic idea of SVM classifier uses this approach, i.e. to choose the hyperplane that has the maximum margin.

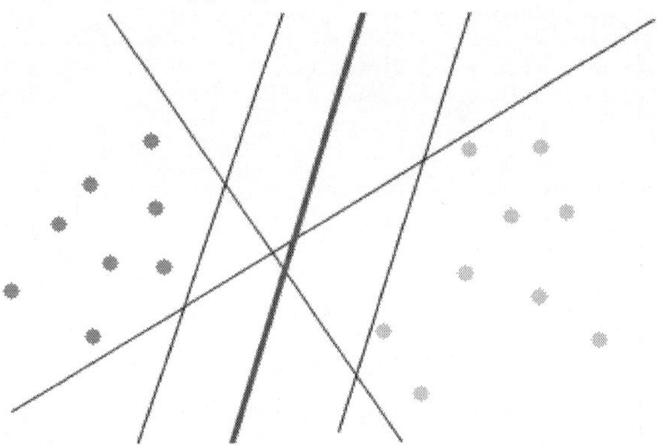

Figure: *Example of separating hyperplanes*

Algorithm

Let D be a classification dataset with n points in a d-dimensional space $D = \{(x_i, y_i)\}$, with $i = 1, 2, ..., n$ and let there be only two class labels such that y_i is either +1 or -1. A hyperplane $h(x)$ gives a linear discriminant function in d dimensions and splits the original space into two half-spaces:

$$h(x) = w^T x + b = w_1 x_1 + w_2 x_2 + ... + w_d x_d + b ,$$

where w is a d-dimensional weight vector and b is a scalar bias. Points on the hyperplane have $h(x) = 0$, i.e. the hyperplane is defined by all points for which $w^T x = -b$.

$$\delta_i = \frac{y_i h(x_i)}{\|w\|}$$

According to, if the dataset is linearly separable, a separating hyperplane can be found such that for all points with label -1, $h(x) < 0$ and for all points labelled +1, $h(x) > 0$. In this case, $h(x)$ serves as a linear classifier or linear discriminant that predicts the class for any point. Moreover, the weight vector w is orthogonal to the hyperplane, therefore giving the direction that is normal to it, whereas the bias b

fixes the offset of the hyperplane in the d-dimensional space. Given a separating hyperplane $h(x) = 0$, it is possible to calculate the distance between each point x_i and the hyperplane by:

The margin of the linear classifier is defined as the minimum distance of all n points to the separating hyperplane.

All points (vectors x^*_i) that achieve this minimum distance are called the support vectors for the linear classifier. In other words, a support vector is a point that lies precisely on the margin of the classifying hyperplane.

$$y_i^* h(x_i^*) = 1$$

In a canonical representation of the hyperplane, for each support vector x^*_i with label y^*_i we have that. Similarly, for any point that is not a support vector, we have that $y_i h(x_i) > 1$, since, by definition, it must be farther from the hyperplane than a support vector. Therefore we have that.

The fundamental idea behind SVMs is to choose the hyperplane with the maximum margin, i.e. the optimal canonical hyperplane. To do this, one needs to find the weight vector w and the bias b that yield the maximum margin among all possible separating hyperplanes, that is, the hyperplane that maximizes $\frac{1}{\|w\|}$. The problem then becomes that of solving a convex minimization problem (notice that instead of maximizing the margin $\frac{1}{\|w\|}$·, one can obtain an equivalent formulation of minimizing $\| w \|$) with linear constraints, as follows:

Objective Function

$$min \frac{\|w\|^2}{2}$$

Linear Constraints

This minimization problem can be solved using the *Lagrange multiplier* method, which introduces a Lagrange multiplier α for each constraint:

This method states that $\alpha_i = 0$ for all points that are at a distance larger than $\frac{1}{\|w\|}$ from the hyperplane, and only for those points that are exactly at the margin, i.e. the support vectors, $\alpha_i > 0$. The weight vector of the classifier is obtained as a linear combination of the support vectors, while the bias is the average of the biases obtained from each support vector.

SVMs can handle linearly non-separable points, where the classes overlap to some extent so that a perfect separation is not possible, by introducing slack variables ε_i for each point x_i in *D*. If $0 \leq \varepsilon_i < 1$, the point is still correctly classified. Otherwise, if $\varepsilon_i > 1$, the point is misclassified. So the goal of the classification becomes that of finding the hyperplane (*w* and *b*) with the maximum margin that also minimizes the sum of slack variables. A methodology similar to that described above is necessary to find the weight vector *w* and the bias *b*.

SVMs can also solve problems with non-linear decision boundaries. The main idea is to map the original *d*-dimensional space into a *d'*-dimensional space (*d'* > *d*), where the points can possibly be linearly separated. Given the original dataset $D = \{x_i, y_i\}$ with $i = 1,...,n$ and the transformation function ϑ, a new dataset is obtained in the transformation space $D_\vartheta = \{\vartheta(x_i), y_i\}$ with $i = 1,...,n$. After the linear decision surface is found in the *d'*-dimensional space, it is mapped back to the non-linear surface in the original *d*-dimensional space. To obtain *w* and *b*, $\vartheta(x)$ needn't be computed in isolation. The only operation required in the transformed space is the inner product $\vartheta(x_i)^T\vartheta(x_j)$, which is defined with the kernel function (*K*) between x_i and x_j. Kernels commonly used with SVMs include:

- the polynomial kernel: $K(x_i,x_j) = (x_i^T x_j + 1)^q$, where q is the degree of the polynomial
- the gaussian kernel: $K(x_i,x_j) = e^{-\frac{\|x_i - x_j\|^2}{2\sigma^2}}$, where σ is the spread or standard deviation.
- the gaussian radial basis function (RBF):

  $$K(x_i,x_j) = e^{-\gamma\|x_i - x_j\|^2}, \gamma \geq 0$$

- the Laplace Radial Basis Function (RBF) kernel:

  $$K(x_i,x_j) = e^{-\gamma\|x_i - x_j\|}, \gamma \geq 0$$

- the hyperbolic tangent kernel:

  $$K(x_i,x_j) = tanh(x_i^T x_j + offset)$$

- the sigmoid kernel:

  $$K(x_i,x_j) = tanh(a x_i^T x_j + offset)$$

- the Bessel function of the first kind kernel:

  $$K(x_i,x_j) = \left(\frac{Bessel_{v+1}^n(\sigma\|x_i - x_j\|)}{(\|x_i - x_j\|)^{-n(v+1)}}\right)$$

- the ANOVA radial basis kernel:

$$K(x_i, x_j) = \left(\sum_{k=1}^{n} e^{-\sigma (x_i^k - x_j^k)^2} \right)^d$$

- the linear splines kernel in one dimension:

$$K(x_i, x_j) = 1 + x_i x_j min(x_i, x_j) - \frac{x_i + x_j}{2} min(x_i, x_j)^2 + \frac{min(x_i, x_j)^3}{3}$$

According to the Gaussian and Laplace RBF and Bessel kernels are general-purpose kernels used when there is no prior knowledge about the data. The linear kernel is useful when dealing with large sparse data vectors as is usually the case in text categorization. The polynomial kernel is popular in image processing, and the sigmoid kernel is mainly used as a proxy for neural networks. The splines and ANOVA RBF kernels typically perform well in regression problems.

Chapter 2

Nature and Characteristics of Geographical Data

"Normal" data can be thought of as the building blocks of information. In modern times, it is often information that is compatible with computers and its inner workings. Geographic data is simply a collection of information that can describe objects and things with relation to space. Often this is done with x,y coordinates or longitudes and latitudes.

So what does this mean for GIS users?

Well geographic data is at the heart of what makes for interesting maps. Without this data, there would be no information that would need to be visualized and analyzed.

ESRI has a great video with a clear example of how geographic data can be visualized. This presentation by Paul Trevillion and Eric Bowman clearly explains how geographic data works and even where you can get it from. It is a bit lengthy but it is very clear in its explanations.

The example used actually talks about movie rentals by zipcode. This really interested me as the data I have been given is organised by zipcodes. The presentation gave me inspiration on how to cartographically treat the data once it is opened up.

Now data can be gotten from many different sources and in the GIS world, data is passed on like heirlooms from person to person. There are sites like the USDA and other government funded sites that maintain data. However, with most data comes the tendency for information about the data to be lost.

Now your probably wondering what's the point about having data about data! The first time I heard this, I thought it was ridiculous too. But when you start working with clunky data, you realise that data about data, also known as metadata, will save your life.

There are things about spatial data or geographic data that make it necessary for you to look up metadata so that all the data you work with line up with one another visually. The first basic concept that you would need to know is that the earth is a 3d object that is being visualized in 2d. With those come distortions you are essentially flattening the globe to see it all. Now, there are different ways of flattening the globe. These different versions are called projections. You need all your data to be in the same projections so that boundaries of your spatial data match up with one another.

Metadata is crucial in determining the validity of your data. Metadata can also help you figure out what type of data you are working with. In geographic data there are two types of data files. You have vector files and raster files. Vector data uses solid lines and curves to visually depict spatial data while raster data uses pixels. Here is a good forum that describes the advantages and disadvantages of each type and when to use each.

Maps and Spatial Information

The main method of identifying and representing the location of geographic features on the landscape is a map. A map is a graphic representation of where features are, explicitly and relative to one another. A map is composed of different geographic features represented as either points, lines, and/or areas. Each feature is defined both by its location in space (with reference to a coordinate system), and by its characteristics (typically referred to as attributes). Quite simply, a map is a model of the real world.

The map legend is the key linking the attributes to the geographic features. Attributes, e.g. such as the species for a forest stand, are typically represented graphically by use of different symbology and/or colour. For GIS, attributes need to be coded in a form in which they can be used for data analysis. This implies loading the attribute data into a database system and *linking* it to the graphic features.

Maps are simply models of the real world. They represent snapshots of the land at a specific map scale. The map legend is the key identifying which features are represented on a map.

For geographic data, often referred to as *spatial data*, features are usually referenced in a coordinate system that models a location on the earth's surface. The coordinate system may be of a variety of types. For natural resource applications the most common are:

➧	geographic coordinates such as latitude and longitude, e.g. 56°27'40" and 116°11'25". These are usually referred by degrees, minutes, and seconds. Geographic coordinates can also be identified as decimal degrees, e.g. 54.65°.
➧	a map projection, e.g. Universe Transverse Mercator (UTM) where coordinates are measured in metres, e.g. 545,000.000 and 6,453,254.000 normally reference to a central meridian. Eastings refer to X coordinates while Northings refer to Y coordinates.
➧	a legal survey description, e.g. Meridian, Township, Range such as the Alberta Township System, e.g. Township 075 Range 10 West of 4th Meridian.

Geographic data is distinguished from *attribute data* in that it is referenced spatially by a coordinate system, e.g. it has a spatial extent. Natural resource applications commonly use a Legal Survey system, e.g. the Alberta Township System (ATS), which identifies feature's locations as being a Meridian, Township, and a Range; or a projection such as the UTM coordinate system which identifies features by an Easting coordinate (X) and a Northing coordinate (Y) in a particular UTM zone.

In the UTM projection the area of the earth between 80 degrees North and 80 degrees South latitude is divided into north-south columns 6 degrees of longitude wide called *zones*. These are numbered 1 to 60 eastward beginning at the 180th meridian. Within each zone the central meridian is given an Easting value of 500,000 metres. The equator is designated as having a Northing value of 0 for northern hemisphere coordinates. Coordinates are recorded relative to the central meridian in metres in a particular zone. The basis of the UTM projection defines that the coordinates are duplicated within each UTM zone. Accordingly, use of the UTM projection is only appropriate for certain spatial extents and scales of data. It is not appropriate to use this projection if your area of interest crosses UTM zone boundaries.

For example, the province of Alberta is located in UTM zones 11 and 12. The central meridian for zone 11 is 117° longitude. The central meridian for zone 12 is 111° longitude. The UTM coordinate system is the most widely used projection in the mapping industry and consequently is becoming an de facto standard for use with geographic information systems. This is particularly true for regional data in Canada. The State Plane coordinate system is widely used in the United States.

Maps are the traditional method of storing and displaying geographic information.

A Map Portrays 3 Kinds of Information about Geographic Features

➧	Location and extent of the feature
➧	Attributes (characteristics) of the feature
➧	Relationship of the feature to other features.

Geography has often been described as the study of why what is where. This description is quite appropriate when considering the three kinds of information that are portrayed by the traditional map;

➧	the location and extent of a feature is identified explicitly by reference to a coordinate system representing the earth's surface. This is where a feature is.
➧	the attributes of a feature describe or characterize the feature. This is what the feature is.
➧	The relationship of a feature to other features is implied from the location and attributes of all features. Relationships can be defined explicitly, e.g. roads connecting towns, regions adjacent to one another, or implicitly, e.g. close to, far from, similar to, etc. Implicit relationships are interpreted according to the knowledge that we have about the natural world. Relationships are described as how or why a feature is.

Characterizing Geographic Features

All geographic features on the earth's surface can be characterized and defined as one of three basic feature types. These are points, lines, and areas.

➧	Point data exists when a feature is associated with a single location in space. Examples of point features include a fire lookout tower, an oil well or gas activity site, and a weather station.
➧	Linear data exists when a feature's location is described by a string of spatial coordinates. Examples of linear data include rivers, roads, pipelines, etc.
➧	Areal data exists when a feature is described by a closed string of spatial coordinates. An area feature is commonly referred to as a polygon. Polygonal data is the most common type of data in natural resource applications. Examples of polygonal data include forest stands, soil classification areas, administrative boundaries, and climate zones. Most polygon data is considered to be homogeneous in nature and thus is consistent throughout.

Every geograpic phenomenon can in principle be represented by either a point, line, and/or an area.

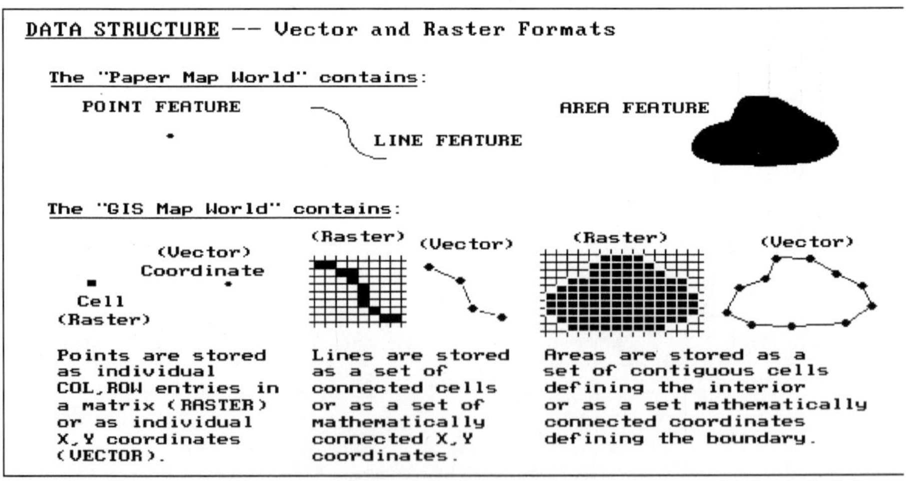

Figure: *GIS Data Structures illustrating the difference between Vector and Raster formats (Adapted from Berry)*

Commonly, an *identifier* accompanies all types of geographic features. This description or identifier is referred to as a label. Labels distinguish geographic features of the same type, e.g. forest stands, from one another. Labels can be in the form of a name, e.g. "Lake Louise", a description, e.g. "WELL" or a unique number, e.g. "123". Forest stand numbers are examples of polygon labels. Each label is unique and provides the mechanism for linking the feature to a set of descriptive characteristics, referred to as attribute data.

It is important to note that geographic features and the symbology used to represent them, e.g. point, line, or polygon, are dependant on the *graphic scale* (*map scale*) of the data. Some features can be represented by point symbology at a small scale, e.g. villages on a 1:1,000,000 map, and by areal symbology at a larger scale, e.g. villages on a 1:10 ,000 map. Accordingly, the accuracy of the feature's location is often *fuzzier* at a smaller scale than a larger scale. The generalization of features is an inherent characteristic of data presented at a smaller scale.

Data can always be generalized to a smaller scale, but detail CANNOT be created !

Remember, as the scale of a map increases, e.g. 1:15,000 to 1:100,000, the relative size of the features decrease and the following may occur:

➡	Some features may disappear, e.g. features such as ponds, hamlets, and lakes, become indistinguishable as a feature and are eliminated.;
➡	Features change from areas to lines or to points, e.g. a village or town represented by a polygon at 1:15,000 may change to point symbology at a 1:100,000 scale.;
➡	Features change in shape, e.g. boundaries become less detailed and more generalized.; and
➡	Some features may appear, e.g. features such as climate zones may be indistinguishable at a large scale (1:15,000) but the full extent of the zone becomes evident at a smaller scale (1:1,000,000).

Accordingly, the use of data from vastly different scales will result in many inconsistencies between the number of features and their type.

The use and comparison of geographic data from vastly different source scales is totally inappropriate and can lead to significant error in geographic data processing.

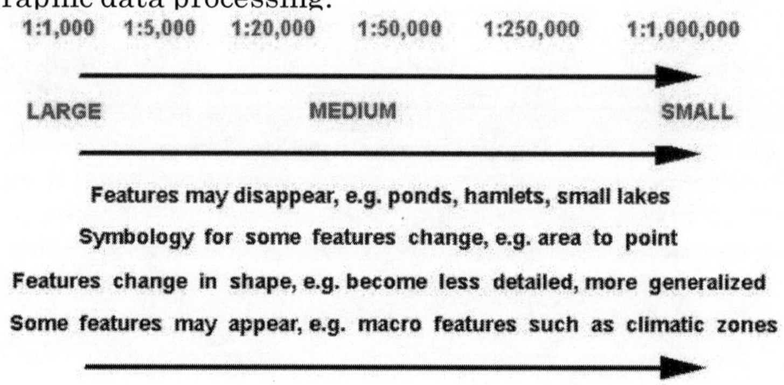

Data Accuracy and Quality

The quality of data sources for GIS processing is becoming an ever increasing concern among GIS application specialists. With the influx of GIS software on the commercial market and the accelerating application of GIS technology to problem solving and decision making roles, the quality and reliability of GIS products is coming under closer scrutiny. Much concern has been raised as to the relative error that may be inherent in GIS processing methodologies. While research is ongoing, and no finite standards have yet been adopted in the commercial GIS marketplace, several practical recommendations have been identified which help to locate possible error sources, and define the quality of data. The following review of data quality focuses on three distinct components, data accuracy, quality, and error.

Accuracy

The fundamental issue with respect to data is *accuracy*. Accuracy is the closeness of results of observations to the true values or values accepted as being true. This implies that observations of most spatial phenomena are usually only considered to estimates of the true value. The difference between observed and true (or accepted as being true) values indicates the accuracy of the observations.

Basically two types of accuracy exist. These are positional and attribute accuracy. Positional accuracyis the expected deviance in the geographic location of an object from its true ground position. This is what we commonly think of when the term accuracy is discussed. There are two components to positional accuracy. These are *relative* and *absolute* accuracy. *Absolute accuracy* concerns the accuracy of data elements with respect to a coordinate scheme, e.g. UTM. *Relative accuracy* concerns the positioning of map features relative to one another.

Often relative accuracy is of greater concern than absolute accuracy. For example, most GIS users can live with the fact that their survey township coordinates do not coincide exactly with the survey fabric, however, the absence of one or two parcels from a tax map can have immediate and costly consequences.

Attribute accuracy is equally as important as positional accuracy. It also reflects estimates of the truth. Interpreting and depicting boundaries and characteristics for forest stands or soil polygons can be exceedingly difficult and subjective. Most resource specialists will attest to this fact. Accordingly, the degree of homogeneity found within such mapped boundaries is not nearly as high in reality as it would appear to be on most maps.

Quality

Quality can simply be defined as the fitness for use for a specific data set. Data that is appropriate for use with one application may not be fit for use with another. It is fully dependant on the scale, accuracy, and extent of the data set, as well as the quality of other data sets to be used. The recent U.S. Spatial Data Transfer Standard (SDTS) identifies five components to data quality definitions. These are :

➡	Lineage
➡	Positional Accuracy
➡	Attribute Accuracy
➡	Logical Consistency
➡	Completeness

Lineage

The lineage of data is concerned with historical and compilation aspects of the data such as the:

➡	source of the data;
➡	content of the data;
➡	data capture specifications;
➡	geographic coverage of the data;
➡	compilation method of the data, e.g. digitizing versus scanned;
➡	ransformation methods applied to the data; and
➡	the use of an pertinent algorithms during compilation, e.g. linear simplification, feature generalization.

Positional Accuracy

The identification of positional accuracy is important. This includes consideration of inherent error (source error) and operational error (introduced error). A more detailed review is provided in the next section.

Attribute Accuracy

Consideration of the accuracy of attributes also helps to define the quality of the data. This quality component concerns the identification of the reliability, or level of purity (homogeneity), in a data set.

Logical Consistency

This component is concerned with determining the faithfulness of the data structure for a data set. This typically involves spatial data inconsistencies such as incorrect line intersections, duplicate lines or boundaries, or gaps in lines. These are referred to as spatial or *topological* errors.

Completeness

The final quality component involves a statement about the completeness of the data set. This includes consideration of *holes* in the data, unclassified areas, and any compilation procedures that may have caused data to be eliminated.

The ease with which geographic data in a GIS can be used at any scale highlights the importance of detailed data quality information. Although a data set may not have a specific scale once it is loaded into the GIS database, it was produced with levels of accuracy and resolution that make it appropriate for use only at certain scales, and in combination with data of similar scales.

Error

Two sources of error, inherent and operational, contribute to the reduction in quality of the products that are generated by geographic information systems. Inherent error is the error present in source documents and data. Operational error is the amount of error produced through the data capture and manipulation functions of a GIS. Possible sources of operational errors include:

➡	Mis-labelling of areas on thematic maps;
➡	misplacement of horizontal (positional) boundaries;
➡	human error in digitizing
➡	classification error;.
➡	GIS algorithm inaccuracies; and
➡	human bias.

While error will always exist in any scientific process, the aim within GIS processing should be to identify existing error in data sources and minimize the amount of error added during processing. Because of cost constraints it is often more appropriate to manage error than attempt to eliminate it. There is a trade-off between reducing the level of error in a data base and the cost to create and maintain the database.

An awareness of the error status of different data sets will allow user to make a subjective statement on the quality and reliability of a product derived from GIS processing.

The validity of any decisions based on a GIS product is directly related to the quality and *reliability rating* of the product.

Depending upon the level of error inherent in the source data, and the error operationally produced through data capture and manipulation, GIS products may possess significant amounts of error.

One of the major problems currently existing within GIS is the aura of accuracy surrounding digital geographic data. Often hardcopy map sources include a map *reliability rating* or *confidence rating* in the map legend. This rating helps the user in determining the *fitness for use* for the map. However, rarely is this information encoded in the digital conversion process.

Often because GIS data is in digital form and can be represented with a high *precision* it is considered to be totally *accurate*. In reality, a *buffer* exists around each feature which represents the actual positional location of the feature. For example, data captured at the 1:20,000 scale commonly has a positional accuracy of +/- 20 metres. This means the actual location of features may vary 20 metres in either direction from the identified position of the feature on the map. Considering that the use of GIS commonly involves the integration of several data sets, usually at different scales and quality, one can easily see how errors can be propagated during processing.

Example of Areas of Uncertainty for Overlaying Data

Several comments and guidelines on the recognition and assessment of error in GIS processing have been promoted in papers on the subject. These are summarized below:

➧	There is a need for developing error statements for data contained within geographic information systems.
➧	The integration of data from different sources and in different original formats (e.g. points, lines, and areas), at different original scales, and possessing inherent errors can yield a product of questionable accuracy.
➧	The accuracy of a GIS-derived product is dependent on characteristics inherent in the source products, and on user requirements, such as scale of the desired output products and the method and resolution of data encoding.
➧	The highest accuracy of any GIS output product can only be as accurate as the least accurate data theme of information involved in the analysis (Newcomer, Szajgin, 1984).
➧	Accuracy of the data decreases as spatial resolution becomes more coarse
➧	As the number of layers in an analysis increases, the number of possible opportunities for error increases.

GIS Data Models and Types

A GIS stores information about the world as a collection of thematic layers that can be linked together by geography. This simple but extremely powerful and versatile concept has proven invaluable for solving many real-world problems from tracking delivery vehicles, to recording details of planning applications, to modelling global atmospheric circulation. The thematic layer approach allows us to organise the complexity of the real world into a simple representation to help facilitate our understanding of natural relationships.

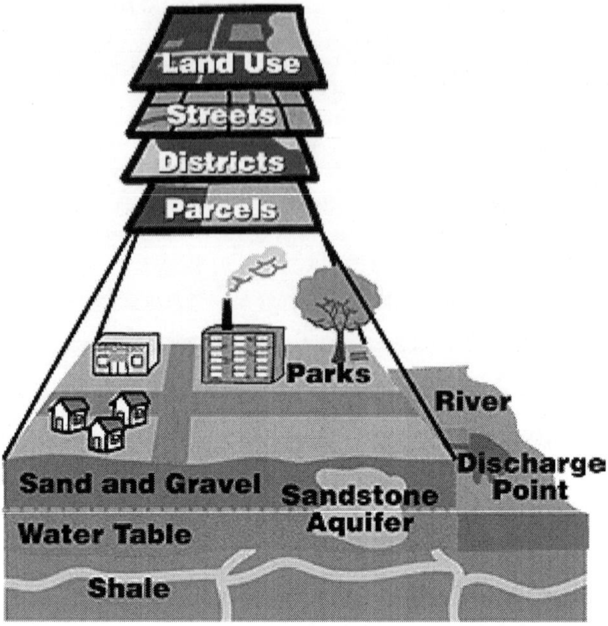

Figure: *The thematic layer approach allows us to organise the complexity of the real world*

GIS Data Types

The basic data type in a GIS reflects traditional data found on a map. Accordingly, GIS technology utilizes two basic types of data. These are:

➡	Spatial data	describes the absolute and relative location of geographic features.
➡	Attribute data	describes characteristics of the spatial features. These characteristics can be quantitative and/or qualitative in nature. Attribute data is often referred to as tabular data.

The coordinate location of a forestry stand would be spatial data, while the characteristics of that forestry stand, e.g. cover group, dominant species, crown closure, height, etc., would be attribute data. Other data types, in particular image and multimedia data, are becoming more prevalent with changing technology. Depending on the specific content of the data, *image data* may be considered either spatial, e.g. photographs, animation, movies, etc., or attribute, e.g. sound, descriptions, narration's, etc.

Spatial Data Models

Traditionally spatial data has been stored and presented in the form of a map. Three basic types of spatial data models have evolved for storing geographic data digitally. These are referred to as:

➡	Vector;
➡	Raster;
➡	Image.

The following diagram reflects the two primary spatial data encoding techniques. These are vector and raster. Image data utilizes techniques very similar to raster data, however typically lacks the internal formats required for analysis and modelling of the data. Images reflect *pictures* or *photographs* of the landscape.

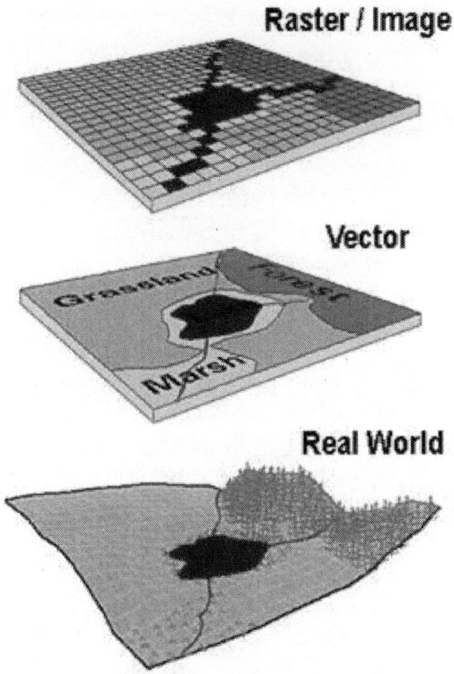

Figure: *Representation of the real world and showing differences in how a vector and a raster GIS will represent this real world.*

Vector Data Formats

All spatial data models are approaches for storing the spatial location of geographic features in a database. Vector storage implies the use of vectors (directional lines) to represent a geographic feature. Vector data is characterized by the use of sequential points or *vertices* to define a linear segment. Each vertex consists of an X coordinate and a Y coordinate.

Vector lines are often referred to as *arcs* and consist of a string of vertices terminated by a *node*. A node is defined as a vertex that starts or ends an arc segment. Point features are defined by one coordinate pair, a vertex. Polygonal features are defined by a set of closed coordinate pairs. In vector representation, the storage of the vertices for each feature is important, as well as the connectivity between features, e.g. the sharing of common vertices where features connect.

Several Different Vector Data Models Exist, However Only Two are Commonly Used in GIS Data Storage.

The most popular method of retaining spatial relationships among features is to explicitly record adjacency information in what is known as the topologic data model. Topology is a mathematical concept that has its basis in the principles of feature adjacency and connectivity.

The topologic data structure is often referred to as an *intelligent data structure* because spatial relationships between geographic features are easily derived when using them. Primarily for this reason the topologic model is the dominant vector data structure currently used in GIS technology. Many of the complex data analysis functions cannot effectively be undertaken without a topologic vector data structure. Topology is reviewed in greater detail later on in the book.

The secondary vector data structure that is common among GIS software is the computer-aided drafting (CAD) data structure. This structure consists of listing elements, not features, defined by strings of vertices, to define geographic features, e.g. points, lines, or areas. There is considerable redundancy with this data model since the boundary segment between two polygons can be stored twice, once for each feature. The CAD structure emerged from the development of computer graphics systems without specific considerations of processing geographic features. Accordingly, since features, e.g. polygons, are self-contained and independent, questions about the adjacency of features can be difficult to answer. The CAD vector model lacks the definition of spatial relationships between features that is defined by the topologic data model.

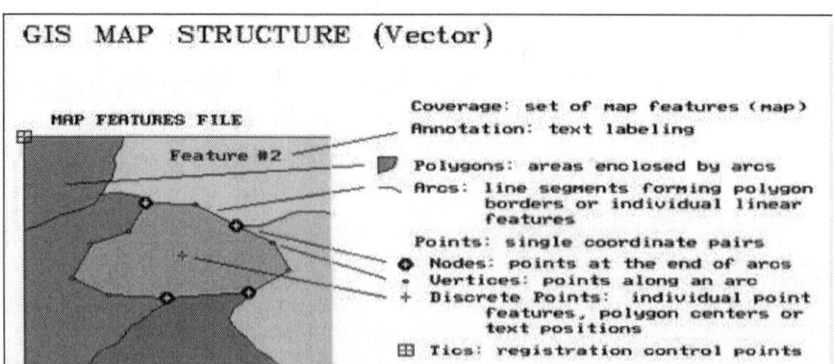

***Figure:** GIS MAP Structure - VECTOR systems (Adapted from Berry)*

Raster Data Formats

Raster data models incorporate the use of a *grid-cell* data structure where the geographic area is divided into cells identified by row and column. This data structure is commonly called *raster*. While the term raster implies a regularly spaced grid other *tessellated* data structures do exist in grid based GIS systems. In particular, the quadtree data structure has found some acceptance as an alternative raster data model.

The size of cells in a tessellated data structure is selected on the basis of the data accuracy and the resolution needed by the user. There is no explicit coding of geographic coordinates required since that is implicit in the layout of the cells. A raster data structure is in fact a matrix where any coordinate can be quickly calculated if the origin point is known, and the size of the grid cells is known. Since grid-cells can be handled as two-dimensional arrays in computer encoding many analytical operations are easy to program. This makes tessellated data structures a popular choice for many GIS software. Topology is not a relevant concept with tessellated structures since adjacency and connectivity are implicit in the location of a particular cell in the data matrix. Several tessellated data structures exist, however only two are commonly used in GIS's. The most popular cell structure is the regularly spaced matrix or *raster* structure. This data structure involves a division of spatial data into regularly spaced cells. Each cell is of the same shape and size. Squares are most commonly utilized.

Since geographic data is rarely distinguished by regularly spaced shapes, cells must be classified as to the most common attribute for the cell. The problem of determining the proper resolution for a particular data layer can be a concern. If one selects too coarse a cell size then data may be overly generalized. If one selects too fine a cell size then too many cells may be created resulting in a large data volume, slower processing times, and a more cumbersome data set. As well, one can imply accuracy greater than that of the original data capture process and this may result in some erroneous results during analysis. As well, since most data is captured in a vector format, e.g. digitizing, data must be converted to the raster data structure. This is called *vector-raster conversion*. Most GIS software allows the user to define the raster grid (cell) size for vector-raster conversion. It is imperative that the original scale, e.g. accuracy, of the data be known prior to conversion. The accuracy of the data, often referred to as the resolution, should determine the cell size of the output raster map during conversion.

Most raster based GIS software requires that the raster cell contain only a single discrete value. Accordingly, a data layer, e.g. forest inventory stands, may be broken down into a series of raster maps, each representing an attribute type, e.g. a species map, a height map, a density map, etc. These are often referred to as *one attribute maps*. This is in contrast to most conventional vector data models that maintain data as *multiple attribute maps*, e.g. forest inventory polygons *linked* to a database table containing all attributes as columns. This basic distinction of raster data storage provides the foundation for quantitative analysis techniques. This is often referred to as *raster or map algebra*. The use of raster data structures allow for sophisticated mathematical modelling processes while vector based systems are often constrained by the capabilities and language of a relational DBMS.

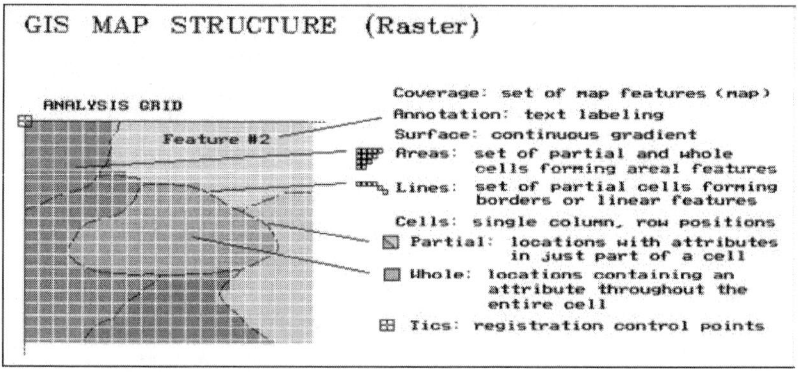

Figure: *GIS MAP Structure - RASTER systems (Adapted from Berry)*

This difference is the major distinguishing factor between vector and raster based GIS software. It is also important to understand that the selection of a particular data structure can provide advantages during the analysis stage. For example, the vector data model does not handle continuous data, e.g. elevation, very well while the raster data model is more ideally suited for this type of analysis. Accordingly, the raster structure does not handle linear data analysis, e.g. shortest path, very well while vector systems do. It is important for the user to understand that there are certain advantages and disadvantages to each data model. The selection of a particular data model, vector or raster, is dependent on the source and type of data, as well as the intended use of the data. Certain analytical procedures require raster data while others are better suited to vector data.

Image Data

Image data is most often used to represent graphic or pictorial data. The term *image* inherently reflects a graphic representation, and in the *GIS world*, differs significantly from raster data. Most often, image data is used to store remotely sensed imagery, e.g. satellite scenes or orthophotos, or ancillary graphics such as photographs, scanned plan documents, etc. Image data is typically used in GIS systems as background display data (if the image has been rectified and georeferenced); or as a graphic attribute. Remote sensing software makes use of image data for image classification and processing. Typically, this data must be converted into a raster format (and perhaps vector) to be used analytically with the GIS.

Image data is typically stored in a variety of de facto industry standard proprietary formats. These often reflect the most popular image processing systems. Other graphic image formats, such as TIFF, GIF, PCX, etc., are used to store ancillary image data. Most GIS software will read such formats and allow you to display this data.

Vector and Raster - Advantages and Disadvantages

There are several advantages and disadvantages for using either the vector or raster data model to store spatial data. These are summarized below.

Vector Data	*Advantages :*
	Data can be represented at its original resolution and form without generalization.
	Graphic output is usually more aesthetically pleasing (traditional cartographic representation);
	Since most data, e.g. hard copy maps, is in vector form no data conversion is required.
	Accurate geographic location of data is maintained.
	Allows for efficient encoding of topology, and as a result more efficient operations that require topological information, e.g. proximity, network analysis.
	Disadvantages:
	The location of each vertex needs to be stored explicitly.
	For effective analysis, vector data must be converted into a topological structure. This is often processing intensive and usually requires extensive data cleaning. As well, topology is static, and any updating or editing of the vector data requires re-building of the topology.
	Algorithms for manipulative and analysis functions are complex and may be processing intensive. Often, this inherently limits the functionality for large data sets, e.g. a large number of features.
	Continuous data, such as elevation data, is not effectively represented in vector form. Usually substantial data generalization or interpolation is required for these data layers.
	Spatial analysis and filtering within polygons is impossible

Raster Data	*Advantages :*
	The geographic location of each cell is implied by its position in the cell matrix. Accordingly, other than an origin point, e.g. bottom left corner, no geographic coordinates are stored.
	Due to the nature of the data storage technique data analysis is usually easy to program and quick to perform.
	The inherent nature of raster maps, e.g. one attribute maps, is ideally suited for mathematical modelling and quantitative analysis.
	Discrete data, e.g. forestry stands, is accommodated equally well as continuous data, e.g. elevation data, and facilitates the integrating of the two data types.
	Grid-cell systems are very compatible with raster-based output devices, e.g. electrostatic plotters, graphic terminals.
	Disadvantages:
	The cell size determines the resolution at which the data is represented.;
	It is especially difficult to adequately represent linear features depending on the cell resolution. Accordingly, network linkages are difficult to establish.
	Processing of associated attribute data may be cumbersome if large amounts of data exists. Raster maps inherently reflect only one attribute or characteristic for an area.
	Since most input data is in vector form, data must undergo vector-to-raster conversion. Besides increased processing requirements this may introduce data integrity concerns due to generalization and choice of inappropriate cell size.
	Most output maps from grid-cell systems do not conform to high-quality cartographic needs.

It is often difficult to compare or rate GIS software that use different data models. Some personal computer (PC) packages utilize vector structures for data input, editing, and display but convert to raster structures for any analysis. Other more comprehensive GIS offerings provide both integrated raster and vector analysis techniques. They allow users to select the data structure appropriate for the analysis requirements. Integrated raster and vector processing capabilities are most desirable and provide the greatest flexibility for data manipulation and analysis.

Attribute Data Models

A separate data model is used to store and maintain attribute data for GIS software. These data models may exist internally within the GIS software, or may be reflected in external commercial Database Management Software (DBMS). A variety of different data models exist for the storage and management of attribute data. The most common are:

➡	Tabular
➡	Hierarchial
➡	Network
➡	Relational
➡	Object Oriented

The tabular model is the manner in which most early GIS software packages stored their attribute data. The next three models are those most commonly implemented in database management systems (DBMS). The object oriented is newer but rapidly gaining in popularity for some applications. A brief review of each model is provided.

Tabular Model

The simple tabular model stores attribute data as sequential data files with fixed formats (or comma delimited for ASCII data), for the location of attribute values in a predefined record structure. This type of data model is outdated in the GIS arena. It lacks any method of checking data integrity, as well as being inefficient with respect to data storage, e.g. limited indexing capability for attributes or records, etc.

Hierarchical Model

The hierarchical database organises data in a *tree* structure. Data is structured downward in a *hierarchy* of tables. Any level in the hierarchy can have unlimited *children*, but any *child* can have only one *parent*. Hierarchial DBMS have not gained any noticeable acceptance for use within GIS. They are oriented for data sets that are very stable, where primary relationships among the data change infrequently or never at all. Also, the limitation on the number of parents that an element may have is not always conducive to actual geographic phenomenon.

Network Model

The network database organises data in a network or *plex* structure. Any column in a plex structure can be linked to any other. Like a tree structure, a plex structure can be described in terms of *parents* and *children*. This model allows for children to have more than one parent.

Network DBMS have not found much more acceptance in GIS than the hierarchical DBMS. They have the same flexibility limitations as hierarchical databases; however, the more powerful structure for representing data relationships allows a more realistic modelling of geographic phenomenon. However, network databases tend to become overly complex too easily. In this regard it is easy to lose control and understanding of the relationships between elements.

Relational Model

The relational database organises data in *tables*. Each table, is identified by a unique table name, and is organised by *rows* and *columns*. Each column within a table also has a unique name. Columns store the values for a specific attribute, e.g. cover group, tree height. Rows represent one record in the table. In a GIS each row is usually linked to a separate spatial feature, e.g. a forestry stand. Accordingly, each row would be comprised of several columns, each column containing a specific value for that geographic feature. The following figure presents a sample table for forest inventory features. This table has 4 rows and 5 columns. The forest stand number would be the *label* for the spatial feature as well as the *primary key* for the database table. This serves as the linkage between the spatial definition of the feature and the attribute data for the feature.

UNIQUE STAND NUMBER	*DOMINANT COVER GROUP*	*AVG. TREE HEIGHT*	*STAND SITE INDEX*	*STAND AGE*
001	DEC	3	G	100
002	DEC-CON	4	M	80
003	DEC-CON	4	M	60
004	CON	4	G	120

Data is often stored in several tables. Tables can be joined or referenced to each other by common columns (relational fields). Usually the common column is an identification number for a selected geographic feature, e.g. a forestry stand polygon number. This identification number acts as the *primary key* for the table. The ability to join tables through use of a common column is the essence of the relational model. Such relational joins are usually ad hoc in nature and form the basis of for querying in a relational GIS product. Unlike the other previously discussed database types, relationships are implicit in the character of the data as opposed to explicit characteristics of the database set up.

The relational database model is the most widely accepted for managing the attributes of geographic data.

There are many different designs of DBMSs, but in GIS the relational design has been the most useful. In the relational design, data are stored conceptually as a collection of tables. Common fields in different tables are used to link them together. This surprisingly simple design has been so widely used primarily because of its flexibility and very wide deployment in applications both within and without GIS.

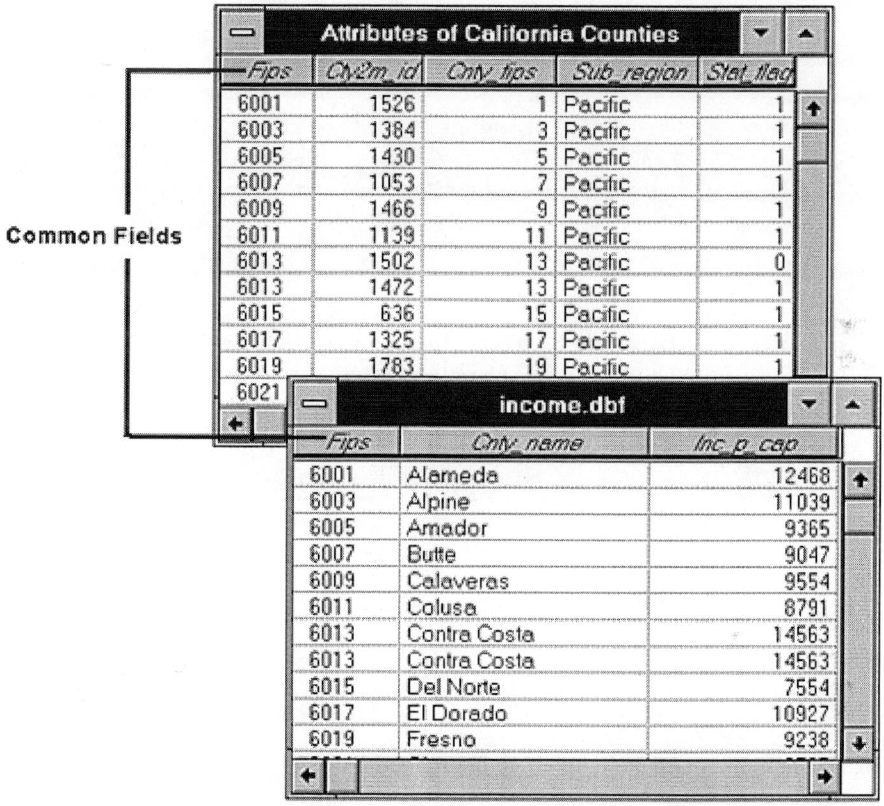

Fips	Cty2m_id	Cnty_fips	Sub_region	Stat_flag
6001	1526	1	Pacific	1
6003	1384	3	Pacific	1
6005	1430	5	Pacific	1
6007	1053	7	Pacific	1
6009	1466	9	Pacific	1
6011	1139	11	Pacific	1
6013	1502	13	Pacific	0
6013	1472	13	Pacific	1
6015	636	15	Pacific	1
6017	1325	17	Pacific	1
6019	1783	19	Pacific	1
6021				

Fips	Cnty_name	Inc_p_cap
6001	Alameda	12468
6003	Alpine	11039
6005	Amador	9365
6007	Butte	9047
6009	Calaveras	9554
6011	Colusa	8791
6013	Contra Costa	14563
6013	Contra Costa	14563
6015	Del Norte	7554
6017	El Dorado	10927
6019	Fresno	9238

Figure: *In the relational design, data are stored conceptually as a collection of tables. Common fields in different tables are used to link them together.*

In fact, most GIS software provides an *internal* relational data model, as well as support for *commercial off-the-shelf* (COTS) relational DBMS'. COTS DBMS' are referred to as *external* DBMS'. This approach supports both users with small data sets, where an internal data model is sufficient, and customers with larger data sets who utilize a DBMS for other corporate data storage requirements. With an external DBMS the GIS software can simply *connect* to the database, and the user can make use of the inherent capabilities of the DBMS. External DBMS' tend to have much more extensive querying and data integrity capabilities than the GIS' internal relational model. The emergence and use of the external DBMS is a trend that has resulted in the proliferation of GIS technology into more traditional data processing environments.

The relational DBMS is attractive because of its:

➡	simplicity in organization and data modelling.
➡	flexibility - data can be manipulated in an ad hoc manner by joining tables.
➡	efficiency of storage - by the proper design of data tables redundant data can be minimized; and
➡	the non-procedural nature - queries on a relational database do not need to take into account the internal organization of the data.

The relational DBMS has emerged as the dominant commercial data management tool in GIS implementation and application.

The following diagram illustrates the basic linkage between a vector spatial data (topologic model) and attributes maintained in a relational database file.

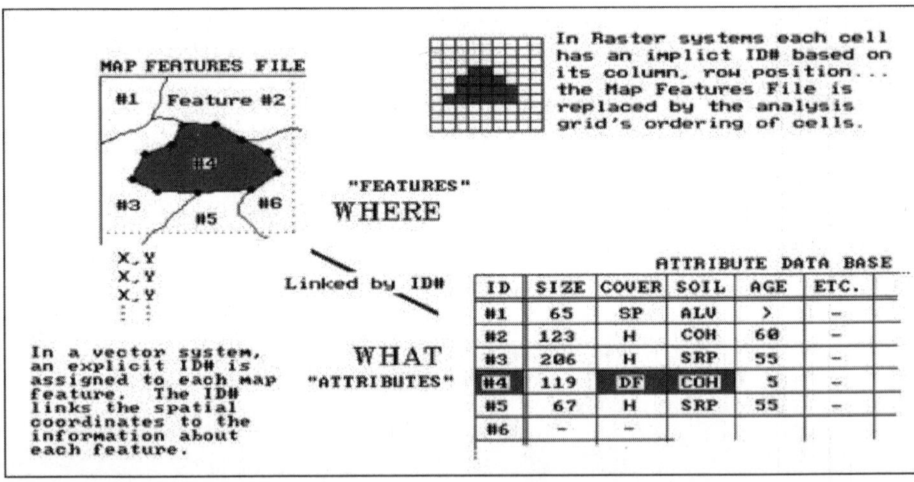

Figure: *Basic linkages between a vector spatial data (topologic model) and attributes maintained in a relational database file*

Object-Oriented Model

The object-oriented database model manages data through *objects.* An object is a collection of data elements and operations that together are considered a single entity. The object-oriented database is a relatively new model. This approach has the attraction that querying is very natural, as features can be bundled together with attributes at the database administrator's discretion. To date, only a few GIS packages are promoting the use of this attribute data model. However, initial impressions indicate that this approach may hold many operational benefits with respect to geographic data processing. Fulfilment of this promise with a commercial GIS product remains to be seen.

Spatial Data Relationships

The nature of spatial data relationships are important to understand within the context of GIS. In particular, the relationship between geographic features is a complex problem in which we are far from understanding in its entirety. This is of concern since the primary role of GIS is the manipulation and analysis of large quantities of spatial data. To date, the accepted theoretical solution is to *topologically structure* spatial data.

It is believed that a topologic data model best reflects the geography of the real world and provides an effective mathematical foundation for encoding spatial relationships, providing a data model for manipulating and analyzing vector based data.

Most GIS software segregate spatial and attribute data into separate data management systems. Most frequently, the topological or raster structure is used to store the spatial data, while the relational database structure is used to store the attribute data. Data from both structures are linked together for use through unique identification numbers, e.g. feature labels and DBMS primary keys. This coupling of spatial features with an attribute record is usually maintained by an internal number assigned by the GIS software. A *label* is required so the user can load the appropriate attribute record for a given geographic feature. Most often a single attribute record is automatically created by the GIS software once a clean topological structure is properly generated. This attribute record normally contains the internal number for the feature, the user's label identifier, the area of the feature, and the perimeter of the feature. Linear features have the length of the feature defined instead of the area.

Topology

The topologic model is often confusing to initial users of GIS. Topology is a mathematical approach that allows us to structure data based on the principles of feature adjacency and feature connectivity. It is in fact the mathematical method used to define spatial relationships. Without a topologic data structure in a vector based GIS most data manipulation and analysis functions would not be practical or feasible.

The most common topological data structure is the *arc/node* data model. This model contains two basic entities, the *arc* and the *node*. The arc is a series of points, joined by straight line segments, that start and end at a node. The node is an intersection point where two or more arcs meet. Nodes also occur at the end of a *dangling* arc, e.g. an arc that does not connect to another arc such as a dead end street. Isolated nodes, not connected to arcs represent point features. A polygon feature is comprised of a closed chain of arcs.

In GIS software the topological definition is commonly stored in a proprietary format. However, most software offerings record the topological definition in three tables. These tables are analogous to relational tables. The three tables represent the different types of features, e.g. point, line, area. A fourth table containing the coordinates is also utilized. The *node table* stores information about the node and the arcs that are connected to it. The *arc table* contains topological information about the arcs. This includes the start and end node, and the polygon to the left and right that the arc is an element of. The *polygon table* defines the arcs that make up each polygon. While arc, node, and polygon terminology is used by most GIS vendors, some also introduce terms such as *edges* and *faces* to define arcs and polygons. This is merely the use of different words to define topological definitions. Do not be confused by this.

Since most input data does not exist in a topological data structure, topology must be *built* with the GIS software. Depending on the data set this can be an CPU intensive and time consuming procedure. This building process involves the creation of the topological tables and the definition of the arc, node, and polygon entities. To properly define the topology there are specific requirements with respect to graphic elements, e.g. no duplicate lines, no gaps in arcs that define polygon features, etc. These requirements are reviewed in the Data Editing section of the book.

The topological model is utilized because it effectively models the relationship of spatial entities. Accordingly, it is well suited for operations such as contiguity and connectivity analyses. Contiguity involves the evaluation of feature adjacency, e.g. features that touch one another, and proximity, e.g. features that are near one another. The primary advantage of the topological model is that spatial analysis can be done without using the coordinate data. Many operations can be done largely, if not entirely, by using the topological definition alone. This is a significant advantage over the CAD or *spaghetti* vector data structure that requires the derivation of spatial relationships from the coordinate data before analysis can be undertaken.

The major disadvantage of the topological data model is its static nature. It can be a time consuming process to properly define the topology depending on the size and complexity of the data set. For example, 2,000 forest stand polygons will require considerably longer to *build* the topology that 2,000 municipal lot boundaries. This is due to the inherent complexity of the features, e.g. lots tend to be rectangular while forest stands are often long and sinuous. This can be a consideration when evaluating the topological building capabilities of GIS software. The static nature of the topological model also implies that every time some editing has occurred, e.g. forest stand boundaries are changed to reflect harvesting or burns, the topology must be rebuilt. The integrity of the topological structure and the DBMS tables containing the attribute data can be a concern here. This is often referred to as *referential integrity*. While topology is the mechanism to ensure integrity with spatial data, referential integrity is the concept of ensuring integrity for both linked topological data and attribute data.

Establishing the Boundaries

A boundary problem in spatial analysis refers to a phenomenon in which geographical patterns are differentiated by the shape and arrangement of boundaries that are drawn for administrative or measurement purposes. This is distinct from and must not be confused with the boundary problem in the philosophy of science that is used as a synonym for the demarcation problem.

Definition

In spatial analysis, four major problems interfere with an accurate estimation of the statistical parameter: the boundary problem, scale problem, pattern problem (or spatial autocorrelation), and modifiable areal unit problem. The boundary problem occurs because of the loss of neighbours in analyses that depend on the values of the neighbours. While geographic phenomena are measured and analyzed within a specific unit, identical spatial data can appear either dispersed or clustered depending on the boundary placed around the data. In analysis with point data, dispersion is evaluated as dependent of the boundary. In analysis with area data, statistics should be interpreted based upon the boundary.

In geographical research, two types of areas are taken into consideration in relation to the boundary: an area surrounded by fixed natural boundaries (e.g., coastlines or streams), outside of which neighbours do not exist, or an area included in a larger region defined by arbitrary artificial boundaries (e.g., an air pollution boundary in modelling studies or an urban boundary in population migration). In an area isolated by the natural boundaries, the spatial process discontinues at the boundaries. In contrast, if a study area is delineated by the artificial boundaries, the process continues beyond the area.

If a spatial process in an area occurs beyond a study area or has an interaction with neighbours outside artificial boundaries, the most common approach is to neglect the influence of the boundaries and assume that the process occurs at the internal area. However, such an approach leads to a significant model misspecification problem.

That is, for measurement or administrative purposes, geographic boundaries are drawn, but the boundaries per se can bring about different spatial patterns in geographic phenomena. It has been reported that the difference in the way of drawing the boundary significantly affects identification of the spatial distribution and estimation of the statistical parameters of the spatial process. The difference is largely based on the fact that spatial processes are generally unbounded or fuzzy-bounded but the processes are expressed in data imposed within boundaries for analysis purposes. Although the boundary problem was discussed in relation to artificial and arbitrary boundaries, the effect of the boundaries also occurs according to natural boundaries as long as it is ignored that properties at sites on the natural boundary such as streams are likely to differ from those at sites within the boundary.

The boundary problem occurs with regard not only to horizontal boundaries but also to vertically drawn boundaries according to delineations of heights or depths (Pineda 1993). For example, biodiversity such as the density of species of plants and animals is high near the surface, so if the identically divided height or depth is used as a spatial unit, it is more likely to find fewer number of the plant and animal species as the height or depth increases.

Types and Examples

By drawing a boundary around a study area, two types of problems in measurement and analysis takes place. The first is an edge effect. This effect originates from the ignorance of interdependences that occur outside the bounded region. Griffith and Griffith and Amrhein highlighted problems according to the edge effect. A typical example is a cross-boundary influence such as cross-border jobs, services and other resources located in a neighbouring municipality.

The second is a shape effect that results from the artificial shape delineated by the boundary. As an illustration of the effect of the artificial shape, point pattern analysis tends to provide higher levels of clustering for the identical point pattern within a unit that is more elongated. Similarly, the shape can influence interaction and flow among spatial entities. For example, the shape can affect the measurement of origin-destination flows since these are often recorded when they cross an artificial boundary. Because of the effect set by the boundary, the shape and area information is used to estimate travel distances from surveys or to locate traffic counters, travel survey stations, or traffic monitoring systems. From the same perspective, Theobald argued that measures of urban sprawl should consider interdependences and interactions with nearby rural areas.

In spatial analysis, the boundary problem has been discussed along with the modifiable areal unit problem (MAUP) inasmuch as MAUP is associated with the arbitrary geographic unit and the unit is defined by the boundary. For administrative purposes, data for policy indicators are usually aggregated within larger units (or enumeration units) such as census tracts, school districts, municipalities and counties. The artificial units serve the purposes of taxation and service provision. For example, municipalities can effectively respond to the need of the public in their jurisdictions. However, in such spatially aggregated units, spatial variations of detailed social variables cannot be identified. The problem is noted when the average degree of a variable and its unequal distribution over space are measured.

Globe

Figure: *A modern raised-relief world globe*

A globe is a three-dimensional scale model of Earth (terrestrial globe or geographical globe) or other celestial body such as a planet or moon. While models can be made of objects with arbitrary or irregular shapes, the term *globe* is used only for models of objects that are approximately spherical. The word "globe" comes from the Latin word *globus*, meaning round mass or sphere. Some terrestrial globes include relief to show mountains and other features on the Earth's surface. There are also globes, called celestial globes or astronomical globes, which are spherical representations of the celestial sphere, showing the apparent positions of the stars and constellations in the sky.

Terrestrial and Planetary

A globe is the only representation of the earth that does not distort either the shape or the size of large features; flat maps are created using a map projection that inevitably introduces an increasing amount of distortion the larger the area that the map shows. A typical scale for a terrestrial globe is roughly 1:40 million. This corresponds to a globe with a circumference of one metre, since the circumference of the real Earth is almost exactly 40 million metres. Sometimes a globe has surface texture showing topography; in these, elevations are exaggerated, otherwise they would be hardly visible. Most modern globes are also imprinted with parallels and meridians, so that one can tell the approximate coordinates of a specific place. Globes may also show the boundaries of countries and their names, a feature that can quickly become out of date, as countries change their name or borders.

Celestial

Celestial globes show the apparent positions of the stars in the sky. They omit the Sun, Moon and planets because the positions of these bodies vary relative to those of the stars, but the ecliptic, along which the Sun moves, is indicated.

A potential issue arises regarding the "handedness" of celestial globes. If the globe is constructed so that the stars are in the positions they actually occupy on the imaginary celestial sphere, then the star field will appear back-to-front on the surface of the globe (all the constellations will appear as their mirror images). This is because the view from Earth, positioned at the centre of the celestial sphere, is of the *inside* of the celestial sphere, whereas the celestial globe is viewed from the *outside*. For this reason, celestial globes may be produced in mirror image, so that at least the constellations appear the "right way round". Some modern celestial globes address this problem by making the surface of the globe transparent. The stars can then be placed in their proper positions and viewed *through* the globe, so that the view is of the inside of the celestial sphere, as it is from Earth.

History

The sphericity of the Earth was established by Greek astronomy in the 3rd century BC, and the earliest terrestrial globe appeared from that period. The earliest known example is the one constructed by Crates of Mallus in Cilicia (now Çukurova in modern-day Turkey), in the mid-2nd century BC.

No terrestrial globes from Antiquity or the Middle Ages have survived. An example of a surviving celestial globe is part of a Hellenistic sculpture, called the Farnese Atlas, surviving in a 2nd-century AD Roman copy in the Naples Archaeological Museum, Italy.

Early terrestrial globes depicting the entirety of the Old World were constructed in the Islamic world. According to David Woodward, one such example was the terrestrial globe introduced to Beijing by the Persian astronomer, Jamal ad-Din, in 1267.

The oldest surviving terrestrial globe is the Erdapfel, created by Martin Behaim in Nuremberg, Germany, in 1492.

Another early globe, the Hunt-Lenox Globe, ca. 1510, is thought to be the source of the phrase *Hic Sunt Dracones*, or "Here be dragons". A similar grapefruit-sized globe made from two halves of an ostrich egg was found in 2012 and is believed to date from 1504. It may be the oldest globe to show the New World. Stefaan Missine, who analyzed the globe for the Washington Map Society journal *Portolan*, said it was "part of an important European collection for decades." After a year of research in which he consulted many experts, Missine concluded the Hunt-Lenox Globe was a copper cast of the egg globe.

A facsimile globe showing America was made by Martin Waldseemueller in 1507. Another "remarkably modern-looking" terrestrial globe of the Earth was constructed by Taqi al-Din at the Istanbul observatory of Taqi al-Din during the 1570s.

The world's first seamless celestial globe was built by Mughal scientists under the patronage of Jahangir.

Globus IMP electro-mechanical devices including five-inch globes have been used in Soviet and Russian spacecraft from 1961 to 2002 as navigation instruments. In 2001, the TMA version of the Soyuz spacecraft replaced this instrument with a virtual globe.

The first terrestrial globe was made in 1492 by Martin Behaim (1459–1537) with help from the painter, Georg Glockendon. Behaim was a German mapmaker, navigator, and merchant. He called his first invention of the globe "Nürnberg Terrestrial Globe." Behaim sailed to different places before he invented the globe. He sailed to Portugal in 1480 as a merchant, to inform King John II about navigation. He then went on a voyage to the coast of West Africa with the Portuguese explorer Diogo Cγo in 1485–1486; during the trip, he discovered the

Congo River. After his return to Nürnberg in 1490, he then began to construct his globe, which seemed inadequate at that time. During the time that Behaim made the globe, there were many blank spots in the map. On the same year that the globe was made, Christopher Columbus landed in a place he thought was the East Indies. But another explorer named Amerigo Vespucci realised that it was the 'gap' on Behaim's globe. The continent of North and South America had been unknown to the people of Europe and Asia until 1492. The new continent was named America by Martin Waldseemueller, a man of the cloth, map and globe maker by trade. Explorers were sent out by emperors, kings, and queens to find and conquer a new land. Upon their return, they would tell their stories of their journey. Soon, the blank spots were disappearing. After years of filling the gaps, the globe has been completed. The globe later then expanded and manufactured to different places. It first started out from Nuremberg, where Behaim started the making, then later to Amsterdam, followed by Venice, Paris, Rome, and London. The United States were last to manufacture in 1810 by James Wilson.

In the 1800s small pocket globes (less than 3 inches) were status symbols for gentlemen and educational toys for rich children.

Chapter 3

Geographical Data: Univarate Methods

"Normal" data can be thought of as the building blocks of information. In modern times, it is often information that is compatible with computers and its inner workings.

Data, information and knowledge are closely related terms, but each has its own role in relation to the other. Data is collected and analysed to create information suitable for making decisions, while knowledge is derived from extensive amounts of experience dealing with information on a subject. For example, the height of Mt. Everest is generally considered to be data. This data may be included in a book along with other data on Mt. Everest to describe the mountain in a manner useful for those who wish to make a decision about the best method to climb it. Using an understanding based on experience climbing mountains to advise persons on the way to reach Mt. Everest's peak may be seen as "knowledge".

That is to say, data is the least abstract, information the next least, and knowledge the most. Data becomes information by interpretation; e.g., the height of Mt. Everest is generally considered as "data", a book on Mt. Everest geological characteristics may be considered as "information", and a report containing practical information on the best way to reach Mt. Everest's peak may be considered as "knowledge".

'Information' bears a diversity of meanings that ranges from everyday to technical. Generally speaking, the concept of information is closely related to notions of constraint, communication, control, data, form, instruction, knowledge, meaning, mental stimulus, pattern, perception, and representation.

Beynon-Davies uses the concept of a sign to distinguish between data and information; data is a series of symbols, while information occurs when the symbols are used to refer to something.

It is people and computers who collect data and impose patterns on it. These patterns are seen as information which can be used to enhance knowledge. These patterns can be interpreted as truth, and are authorized as aesthetic and ethical criteria. Events that leave behind perceivable physical or virtual remains can be traced back through data. Marks are no longer considered data once the link between the mark and observation is broken.

Mechanical computing devices are classified according to the means by which they represent data. An analog computer represents a datum as a voltage, distance, position, or other physical quantity. A digital computer represents a piece of data as a sequence of symbols drawn from a fixed alphabet. The most common digital computers use a binary alphabet, that is, an alphabet of two characters, typically denoted "0" and "1". More familiar representations, such as numbers or letters, are then constructed from the binary alphabet.

Some special forms of data are distinguished. A computer programme is a collection of data, which can be interpreted as instructions. Most computer languages make a distinction between programmes and the other data on which programmes operate, but in some languages, notably Lisp and similar languages, programmes are essentially indistinguishable from other data. It is also useful to distinguish metadata, that is, a description of other data. A similar yet earlier term for metadata is "ancillary data." The prototypical example of metadata is the library catalogue, which is a description of the contents of books.

Geographic data is simply a collection of information that can describe objects and things with relation to space. Often this is done with x,y coordinates or longitudes and latitudes. Geographic data is at the heart of what makes for interesting maps. Without this data, there would be no information that would need to be visualized and analysed.

Geographic data can be used to determine a variety of characteristics of a population. Information gleaned from research using geographic data enables you to compare basic details such as economic status, average age, and ethnic diversity in different areas of the country.

Though data is also increasingly used in other fields it has been suggested that the highly interpretive nature of them might be at odds with the ethos of data as "given". Peter Checkland introduced the term *capta*) to distinguish between an immense number of possible data and a sub-set of them, to which attention is oriented. Johanna Drucker has argued that since the humanities affirm knowledge production as "situated, partial, and constitutive," using *data* may introduce assumptions that are counterproductive, for example that phenomena are discrete or are observer-independent. The term *capta,* which emphasizes the act of observation as constitutive, is offered as an alternative to *data* for visual representations in the humanities.

Data Model (GIS)

A data model in geographic information systems is a mathematical construct for representing geographic objects or surfaces as data. For example, the vector data model represents geography as collections of points, lines, and polygons; the raster data model represent geography as cell matrices that store numeric values; and the TIN data model represents geography as sets of contiguous, nonoverlapping triangles.

Representing three-dimensional map information

There are two approaches for representing three-dimensional map information, and for managing it in the data model.

Vector-based Stack-unit

Vector-based stack-unit maps depict the vertical succession of geologic units to a specified depth (here, the base of the block diagram). This mapping approach characterizes the vertical variations of physical properties in each 3-D map unit. In this example, an alluvial deposit (unit "a") overlies glacial till (unit "t"), and the stack-unit labelled "a/t" indicates that relationship, whereas the unit "t" indicates that glacial till extends down to the specified depth. The stack-unit's occurrence (the map unit's outcrop), geometry (the map unit's boundaries), and descriptors (the physical properties of the geologic units included in the stack-unit) are managed as they are for a typical 2-D geologic map.

Raster-based Stacked Surfaces

Raster-based stacked surfaces depict the surface of each buried geologic unit, and can accommodate data on lateral variations of physical properties. In this example from Soller and others (1999), the upper surface of each buried geologic unit was represented in raster format as an ArcInfo Grid file. The middle grid is the uppermost surface of an economically important aquifer, the Mahomet Sand, which fills a pre- and inter-glacial valley carved into the bedrock surface. Each geologic unit in raster format can be managed in the data model, in a manner not dissimilar from that shown for the stack-unit map. The

Mahomet Sand is continuous in this area, and represents one occurrence of this unit in the data model. Each raster, or pixel, on the Mahomet Sand surface has a set of map coordinates that are recorded in a GIS (in the data model bin that is labelled "Pixel coordinates", which is the raster corollary of the "Geometry" bin for vector map data). Each pixel can have a unique set of descriptive information, such as surface elevation, unit thickness, lithology, transmissivity, etc.).

Formats for Geographic Data

Data are nothing but references to observations and measurements related to real-world or imaginary entities and phenomena. There are three fundamental means for organizing spatial data: Tables, Vector Feature Classes, and Raster Layers. These basic data provide a predictable means of organization, —schema—, that permit our tools to exchange information and to engage information with operations and to discover associations concerning information from different data sources.

Related Documents

- ArcMap 101: Introduction to GIS Data and Portrayal provides a tour through an arcmap dataset with many different types of data and metadata.
- Sources of Geographic Data Discusses sources and strategies for finding data that can be used in GIS.
- Understanding GIS Data, Referencing Systems and Metadata. discusses a critical mindset for understanding the quality and fitness of data for use in particular situations.
- ArcMap 10 Help on Data Formats Supported in ArcMap

As we look more deeply into these structures, we will see that in each of the formal categories of data structure: Tabular, Vector & Raster, there are many choices in terms of the Technical Implementation for the way that they are encoded. Each of these is manifest in different data formats. In choosing one encoding scheme over another, we are often trade one virtue for another. For example we may choose to represent our tabular information as a text file or in excel, we may choose to represent vector data in a ESRI Shape File, a AutoDesk DWG file or in a Geodatabase Feature Class. Raster data may be stored an exchanged in any number of popular formats, TIF, JPEG, SID, JP2, GIF, or we may use a GIS specific format, such as GeoTIFF, ArcInfo GRID or Imagine .IMG. The following provides a brief summary of comparitive advantages and disadvantages of one choice of data structure over another.

Why are there so many choices?? Will humans ever agree find one set of data formats that we will use for everything? At the heart of the problem lies the tension between Stability (a strong standard is one that does not change) and Innovation (people keep coming up with better ways of doing things — and new things to do that challenge the capabilities of new standards.) Likely as not, this is not going to change, and therefore data wranglers are going to have to keep learning new things about formats for encoding our spatially referenced observations.

Old and Stable vs New and Innovative may be thought of as one dimension upon which any data format might be placed. Here are several more considerations that may come into play when evaluating a particular data format.

- Simple, file based versus Complex database structures
- Open, community regulated formats versus Proprietary formats.
- Supporting georeferenced coordinate systems versus Not Georeferenced.
- Semantic Depth: Shallow and fixed versus Flexible and deep.
- Topology: Simple versus Sophisticated
- Desktop-oriented versus Client-Server based

Tables

Plain Text, Comma Delimited files. The ultimate in Simple, Open data formats. Text files are very shallow in terms of their ability to assign logical types to specific data fields, but software like ArcGIS will try to guess whether a specific column of text is referring to Text, Numbers, or Dates. Sometimes the software gets it wrong, such as treating zipcodes as numbers.

- DBase Format. Dbase tables are an open format for exchanging tabular information that offers a deeper capacity to assign specific logical datatypes to fields. One drawback of dbase tables is that column names can't be more than 9 characters. Dbase is a static specification, which makes it stable, and used in may tools. Yet it seems to be being abandoned by Microsoft.
- Excel Worksheets Excel is a wonderrful tool that offers a very deep ability to represent tabular information. The fact that one column may be specified to be a dynamic function of other columns is an example of this depth. A drawback of excel is that it is much more complex than text or dbase as a file format, and should not be expected to be transferrable to a lot of other applications — especially as this format may change at any time at the whim of its owner.

- Desktop-Oriented Database Formats Fro example, Microsoft Access is a tool that lets you make very complex arrangements of tables. From the filesystem, the whole complex of tables is represented as a single file. These database applications offer advantages in terms of the size of tables that may be supported, To get inside one of these things a software developer must pay licensing fees to Microsoft for the special programmes involved.
- Enterprise Scale, Server-Oriented Databases IN an organization where many people and applications may need to access and even update the latest version of a table, we find server-based database management systems like Oracle, MYSQL, Postgres Microsoft SQL, and others. These Relational Database Management Systems (RDBMS) can handle (theoretically) unlimited amounts of data and serve it very efficiently to applications that make requests according to standardized protocols. These tools can also manage passwords and permissions and grant privileges to particular users. It is worth noting that the International Standards Organization, (ISO) has specific standards for the fundamental logical data teypes that must be supported in a relational database management system, and the protocols that they must support for exchanging data with applications that requst it. Even if the technical data format used to store the data on disks is completely incomprehensible, these systems are considered to be open and stable, to the degree that they support the ISO specification. For archival purposes, dumps must be made in open format exchange files.

Vector Data Formats

With the exception of the popular CAD formats, .dwg and .dxf, vector formats supported in GIS are extensions of the tabular types discussed above. In effect, Points, Lines and Polygons merely extend the range of datatypes and associated logic traditionally offered in tables The ISO standards for logical datatypes were extended to handle spatial data: points, lines, polygons, and surfaces; in the mid 1990s. In the world of ArcMap, individual datasets representing collections of objects each having the same type of feature are known as Feature Classes.

- Text Files By including a numeric fields for X and Y coordinates, you can associate references to point entities in a text file. Note that the way that these coordinates relate to places on the planet requires some metadata that specifies the specific Spatial Referencing System that is employed.

- DWG and DXF Formats These are proprietary formats of the AutoDesk corporation, though they are considered to be open and documented, there is nothing preventing AutoDesk from changing these. The ability to attach semantic information to geometry in these formats is very limited and fixed. This leads people to develop elaborate schema that distinquish among entities according to their layer and colour. These formats are also very flexible in terms of the ability to mix all sorts of geometry types in a single dataset, which canmake it difficult to create models that function predictably from a GIS perspective. ArcGIS can read DWG and DXF directly and can export to CAD format files and attach georeferencing metadata to them.
- Shape Files Shape files are more or less an extension of DBase data format to handle spatial data types. The Shapefile format is owned by ESRI, but it is openly documented and considered to be a stable exchange format, for now. Shape files can cary their spatial referencing information in an associated .prj file, however, shape files created before 2003(?) will not have a prj file and will therefore not cooperate with ArcGIS on-the-fly projection.
- File Based, and Personal Geodatabases ArcGIS has extended the Microsoft Access data format .mdb files for handling spatial data types. This gives the flexibility to have long field names and large datasets, and complex relationships among tables. One really useful thing about geodatabses is that the geometric properties of features, i.e. their area and perimeter is automatically updated when the geometry is changed. In version 9.0, ESRI introduced its own File Geodatabase format for geodatabases that is not tied to the microsoft database format.
- Enterprise-Scale Geodatabases An Enterprise Scale Geodatabase is a means of using server-based relational database management tools, like Oracle of Postgres to store feature classes. The nature of relational database management systems is such that these datasets have no limit in terms of size.
- Web Feature Services (WFS) is a means of sending styled points lines and polygons to web clients (not supported in simple web browsers. Open protocols allow GIS operations to be performed on these features. A Transactional Web Feature Service (WFST)even allows clients to edit features through the web. Web Feature Services are stable community-based standards of the OGC..

Raster Image Formats

While Tabular data structures allow us to distinquish and form associations among different classes of discrete entities, Raster Images provide containers for representations of locations. Locations are identified by cells or pixels, that can be associated with attributes. From a perspective of GIS, there are a couple of important aspects of rasters:

- Do they carry georeferencing information, linking the cell locations to specific coordinates and or relationship with places on the globe? ArcMap can Georeference any sort of image using its own scheme of world files and aux.xml files that can be associated with an image. But some image formats contain their own internal georeferencing information.
- What is the bit depth that is supported by the attribute referencing system.
- Does the raster format support compression, and how loss is it?
- Bitmap Images Each cell can have one of two values: 1, or 0, Black or White.
- Colour Mapped Images, 8-bit GIF, and PNG images typically support 256 different distinctions which can be mapped to specific colours, or transparancy.
- 8-Bit Gray Scale Images Each pixel is assigned an attribute from 0-255, representing a gradient.
- True-Colour, Multi-Band Images TIFF is a format that integrates three channels of 8 bit intensity that is useful for representing mixtures of Red, Green and Blue. Because of these 3 8-bit channels, supporting 16,777,216 distinct combinations of colour. There is a GeoTIFF format which supports internal georefencing.
- Compressed Image Formats Jpg is an example of a means of compressing image information. These images have a 24 bit depth, but depending on the compression, the edges of things, the actual assignment of data to pixels ins not necessarily predictable, even if it looks right.
- Wavelet Compression SID files and JPEG2000 format files use a progressive compression method that responds to how close you are zoomed in. The compression is much better than plain jpeg and both of these formats carry internal georeferencing capabilities — though they are not always used.

- Deeper Raster Formats When you have a raster GIS dataset that requires a continuous range of attributes that extends beyond 256 values in a single channel, you need a deeper raster structure. These are provided by the proprietary ArcInfo GRID format or the simpler and more open Imagine IMG format.
- Image Map Services (WMS) are a means of sending georeferenced images to a web browser. These images may be pre-tiled or composed on the fly — effectively making a map the server, taking an image of it and sending it to the browser. Google Maps is an example of an image map services. The open Geospatial Consortium has a very stable and popular specification for these known as the Web Map Service (WMS). These may be viewed in ordinary web browsers.

Data Models and Schema

Typical GIS databases that we collect frm sources are fairly elemental, consisting of discrete collections of vector features or individual raster layers. However, it is also useful to consider how complexes of feature classes and rasters can be organized to make data models that are coherent in terms of the relationships among features and potenmtially also engaged with rasters. Higher order data collections that operate this way are thought of as Schema in the sense of their abstract organization, or as Data Models when they are implemented and used. An advantage of thinking of schema in this way is that toolkits may be developed that develop inferences and perform experiments involving the consitiution of elements and relationships among them.

Most schema that we make are relatively ad-hoc. However a very important movement in the field of GIS an other information management endeavours is to develop elaborate schema with organizations of people who will be better able to exchange very deep information and tools. A hallmark of this movement is the use of XML (Extensible Markup Language) — which is a sort of meta schema. The development XML schema by communities of interest is driving a revolution in collaborative information models that can be systematically exchanged — known as the Semantic Web. A branch of the open source movement, these efforts usually involve cross-disciplinary collaborations in which participants understand that the development of a shared language will enhance their niche in the ever-more diverse information ecology. There are several non-profit collaborations that are very active in developing very useful schema.

Univariate

In mathematics, univariate refers to an expression, equation, function or polynomial of only one variable. Objects of any of these types but involving more than one variable may be called multivariate. In some cases the distinction between the univariate and multivariate cases is fundamental; for example, the fundamental theorem of algebra and Euclid's algorithm for polynomials are fundamental properties of univariate polynomials that can not be generalized to multivariate polynomials.

The term is commonly used in statistics to distinguish a distribution of one variable from a distribution of several variables, although it can be applied in other ways as well. For example, univariate data are composed of a single scalar component. In time series analysis, the term is applied with a whole time series as the object referred to: thus a univariate time series refers to the set of values over time of a single quantity. Correspondingly, a "multivariate time series" refers to the changing values over time of several quantities. Thus there is a minor conflict of terminology since the values within a univariate time series may be treated using certain types of multivariate statistical analyses and may be represented using multivariate distributions.

Univariate Analysis

Univariate analysis is the simplest form of quantitative (statistical) analysis. The analysis is carried out with the description of a single variable in terms of the applicable unit of analysis. For example, if the variable "age" was the subject of the analysis, the researcher would look at how many subjects fall into given age attribute categories.

Univariate analysis contrasts with bivariate analysis – the analysis of two variables simultaneously – or multivariable analysis – the analysis of multiple variables simultaneously. Univariate analysis is commonly used in the first, descriptive stages of research, before being supplemented by more advanced, inferential bivariate or multivariate analysis.

Methods

A basic way of presenting univariate data is to create a frequency distribution of the individual cases, which involves presenting the number of cases in the sample that fall into each category of values of the variable. This can be done in a table format or with a bar chart or a similar form of graphical representation. A sample distribution table is presented below, showing the frequency distribution for a variable "age".

Age range	Number of cases	Percent
under 18	10	5
18–29	50	25
29–45	40	20
45–65	40	20
over 65	60	30
Valid cases: 200 Missing cases: 0		

In addition to frequency distribution, univariate analysis commonly involves reporting measures of central tendency (location). This involves describing the way in which quantitative data tend to cluster around some value. In univariate analysis, the measure of central tendency is an average of a set of measurements, the word "average" being variously construed as (arithmetic) mean, median, mode or another measure of location, depending on the context. For a categorical variable, such as preferred brand of cereal, only the mode can serve this purpose. For a variable measured on an interval scale, such as temperature on the Celsius scale, or on a ratio scale, such as temperature on the Kelvin scale, the median or mean can also be used.

Issues in Univariate Forecasting

Univariate modelling just involve one variable i.e. a set of time series data. The series has its own pattern whether trend, seasonal, cycle or irregular. The identification of components is important towards determination of suitable technique.

Thematic Map

A thematic map is a type of map especially designed to show a particular theme connected with a specific geographic area. These maps "can portray physical, social, political, cultural, economic, sociological, agricultural, or any other aspects of a city, state, region, nation, or continent".

A thematic map is a map that focuses on a specific theme or subject area, whereas in a general map the variety of phenomena—geological, geographical, political—regularly appear together. The contrast between them lies in the fact that thematic maps use the base data, such as coastlines, boundaries and places, only my points of reference for the phenomenon being mapped. General maps portray the base data, such as landforms, lines of transportation, settlements, and political boundaries, for their own sake.

Thematic maps emphasize spatial variation of one or a small number of geographic distributions. These distributions may be physical phenomena such as climate or human characteristics such as population density and health issues. Barbara Petchenik described the difference as "in place, about space." While general reference maps show where something is in space, thematic maps tell a story about that place (e.g., city map).

Thematic map are sometimes referred to as graphic essays that portray spatial variations and interrelationships of geographical distributions. Location, of course, is important to provide a reference base of where selected phenomena are occurring.

History

An important cartographic element preceding thematic mapping was the development of accurate base maps. Improvements in accuracy proceeded at a gradual pace, and even until the mid-17th century, general maps were usually of poor quality. Still, base maps around this time were good enough to display appropriate information, allowing for the first thematic maps to come into being.

One of the earliest thematic maps was a map entitled *Designatio orbis christiani* (1607) by Jodocus Hondius showing the dispersion of major religions, using map symbols in the French edition of his *Atlas Minor* (1607). This was soon followed by a thematic globe (in the form of a six-gore map) showing the same subject, using Hondius' symbols, by Franciscus Haraeus, entitled: *Novus typus orbis ipsus globus, ex Analemmate Ptolomaei diductus* (1614)

An early contributor to thematic mapping in England was the English astronomer Edmond Halley (1656–1742). His first significant cartographic contribution was a star chart of the constellation of the Southern Hemisphere, made during his stay on St. Helena and published on 1686. In that same year he also published his first terrestrial map in an article about trade winds, and this map is called the first meteorological chart. In 1701 he published the "New and Correct Chart Shewing the Variations of the Compass".

Another example of early thematic mapping comes from London physician John Snow. Though disease had been mapped thematically, Snow's cholera map in 1854 is the best known example of using thematic maps for analysis. Essentially, his technique and methodology anticipate principles of a geographic information system (GIS). Starting with an accurate base map of a London neighbourhood which included streets and water pump locations, Snow mapped out the incidents of cholera death. The emerging pattern centred around one particular pump on Broad Street. At Snow's request, the handle of the pump was removed, and new cholera cases ceased almost at once. Further investigation of the area revealed the Broad Street pump was near a cesspit under the home of the outbreak's first cholera victim.

Another 19th century example of thematic maps, according to Friendly (2008), was the earliest known choropleth map in 1826 created by Charles Dupin. Based on this work Louis-Leger Gauthier (1815–1881) developed the population contour map, a map that shows the population density by contours or isolines.

Uses of Thematic Maps

Thematic maps serve three primary purposes.

- First, they provide specific information about particular locations.
- Second, they provide general information about spatial patterns.
- Third, they can be used to compare patterns on two or more maps.

Common examples are maps of demographic data such as population density. When designing a thematic map, cartographers must balance a number of factors in order to effectively represent the data. Besides spatial accuracy, and aesthetics, quirks of human visual perception and the presentation format must be taken into account.

In addition, the audience is of equal importance. Who will "read" the thematic map and for what purpose helps define how it should be designed. A political scientist might prefer having information mapped within clearly delineated county boundaries (choropleth maps). A state biologist could certainly benefit from county boundaries being on a map, but nature seldom falls into such smooth, man-made delineations. In which case, a dasymetric map charts the desired information underneath a transparent county boundary map for easy location referencing.

Data Terminology

A thematic map is univariate if the non-location data is all of the same kind. Population density, cancer rates, and annual rainfall are three examples of univariate data. Bivariate mapping shows the geographical distribution of two distinct sets of data. For example, a map showing both rainfall and cancer rates may be used to explore a possible correlation between the two phenomena.

More than two sets of data leads to multivariate mapping. For example, a single map might show population density in addition to annual rainfall and cancer rates.

Methods of Thematic Mapping

Cartographers use many methods to create thematic maps, but five techniques are especially noted.

Choropleth

Choropleth mapping shows statistical data aggregated over predefined regions, such as counties or states, by colouring or shading these regions. For example, countries with higher rates of infant mortality might appear darker on a choropleth map. This technique assumes a relatively even distribution of the measured phenomenon within each region. Generally speaking, differences in hue are used to indicate qualitative differences, such as land use, while differences in saturation or lightness are used to indicate quantitative differences, such as population.

Proportional Symbol

The proportional symbol technique uses symbols of different sizes to represent data associated with different areas or locations within the map. For example, a disc may be shown at the location of each city in a map, with the area of the disc being proportional to the population of the city.

Isarithmic or Isopleth

Isarithmic maps, also known as contour maps or isopleth maps depict smooth continuous phenomena such as precipitation or elevation. Each line-bounded area on this type of map represents a region with the same value. For example, on an elevation map, each elevation line indicates an area at the listed elevation. An Isarithmic map is a planimetric graphic representation of a 3-D surface. Isarithmic mapping requires 3-D thinking for surfaces that vary spatially.

Dot

A dot distribution map might be used to locate each occurrence of a phenomenon, as in Dr. Snow's map where each dot represented one death due to cholera. Where appropriate, a dot may indicate any number of entities, for example, one dot for every 100 voters.

Dasymetric

A dasymetric map is an alternative to a choropleth map. As with a choropleth map, data are collected by enumeration units. But instead of mapping the data so that the region appears uniform, *ancillary information* is used to model internal distribution of the phenomenon. For example, population density will be much lower in forested area than urbanized area, so in a common operation, land cover data (forest, water, grassland, urbanization) may be used to model the distribution of population reported by census enumeration unit such as a tract or county.

Methods of Generating Surfaces in Environmental GIS Applications

In environmental GIS applications, generating surfaces is a frequently imposed requirement in the early stage of analyses. Example of surfaces include topographic surfaces, bedrock surfaces, groundwater tables, and contamination plumes for the specific media. Surfaces like these provide critical information for subsequent analyses which involves subjects such as groundwater modelling, saturated thickness, subsurface distribution of contamination, and volume estimation.

Data for generating these surfaces are usually collected though field sampling and surveying. Because of high cost and limited resource, data collection can be conducted only in selected point locations with limited numbers. In order to generate a continuous surface of a property (i.e. groundwater table), some kind of interpolation method has to be used to estimate surface values at those locations where no samples or measurements were taken.

ArcInfo offers four interpolation methods: trend surface (TREND), inverse distance weighted (IDW), triangulation (CREATETIN), and kriging (KRIGING). The triangulation method generates surfaces represented by irregularly spaced points (TIN). The other three methods generate surfaces represented by equally spaced data points (GRID). In terms of interpolation, each of these method has its own advantages and disadvantages. No one method works universally as the best for any data set. The intent of this paper is to give a comparative evaluation of these interpolation methods and provide some insights on how these methods should used properly to generate surfaces.

Data Evaluation and Preparation

This is always the first and an important step for generating any surfaces since different interpolation methods work best for different data set. A good understanding of the data set is an essential ingredient of good interpolation. The time taken to explore, understand, and describe the data set should be amply regarded though quick adoption of the most suitable interpolation method.

Data evaluation consists of two parts: quality evaluation and statistical evaluation. Quality evaluation is to ensure that data are accurate and representative. For example, well data for generating groundwater table should be based on water level readings on same date and from the same type of wells (such as shallow monitoring wells). In order to accurately represent normal groundwater table, sampling date should be carefully chosen to reflect the normal weather conditions.

Chemical and radiological data should be verified and validated before they are used to generate concentration surfaces. If these data are reported from several laboratories, special check should be made to eliminate the differences among laboratories. It is possible that the same sample may have different concentration or activity values because different laboratories may use different instruments or analytical procedures. Another issue is related to the use of analysed values that are below the detection limits. In most cases, detection limits are used as the substituted concentration or activity values for samples below detection limits. However, other statistically defined values can also used.

Spatial attention should be given to the abnormal values appeared in the data set since these value will seriously distort the interpolation process. The best way to detect these value is though map or visual display of the data set. Abnormal values usually appear as "spatial outliner", i.e. an extreme low value are surrounded by high values or visa versa. Map display also provides additional insights about the data set: how samples are spatially distributed (uniform or cluster), and what is the general pattern and trend (high values and low values) associated with the data set.

Statistical evaluation is to understand statistical properties associated with the dada set. These properties may include distribution, location, spread, and shape of the data set. There are many tools available in the univariate description statistics which can be used to describe these properties. For example, the frequency table and corresponding historgram can be used to describe how often observed values fall within certain intervals or classes. Probability plot can be used to determine how close the distribution of the data set is to a Gaussian or normal distribution.

Summary statistics provide a valuable summary of information contained in the historgram. It includes three groups: measures of location, measures of spread, and measures of shape. Measures of location describe where various parts of the distribution lie. The centre of distribution is described by the mean, the median, and the mode. The location of other part of the distribution are given by various quantiles. Measures of spread describe the variability. This group includes the variance, the stand deviation, and the interquarterile range. Measures of shape include coefficient of skewness and coefficient of variation. Coefficient of skewness provides information on the symmetry while coefficient of variation gives information on the length of the tail for certain types of distributions.

Interpolation Method

A frequently asked question is what is the best interpolation method for any data set. The answer to this question is none. There are many interpolation methods available. Each of these method works best for a particular data set because of its inherent assumptions and algorithm design for estimation. For a given data set, different interpolation methods may work best for different study objectives (i.e. smooth surface vs. accurate surface). Although neighbourhood-based linear interpolation with blended gradients may seem to be a more general and flexible method, there is still no interpolation method which will guarantee the best results for all data sets. With this statement in mind, selection of a particular method should depend upon characteristics of data set as well as study objectives.

The following discussions focus on four interpolation methods which can be easily accessed by ArcInfo users: trend surface, inverse distance weighted, triangulation, and kriging. Among these four methods, trend surface acts as global interpolation method while other three method function as local interpolation methods. With global interpolation methods, all observational points within a study area are utilized to estimate the value at a new point. With local interpolation methods, only a subset of observational points located near the new point are used to estimate the value at this point.

1. *Trend Surface:* The idea behind the trend surface interpolation is to fit a least-squares surface to observational data points by using polynomial regression. The advantage of this method is that it is superficially easy to understand, at least with respect to the way the surfaces are estimated. It can be used to show broad features of the observational data points, such as the overall flow direction of groundwater.

   The disadvantage is that this method is highly affected by the extreme values and uneven distribution of observational data points. The problem is further complicated by the fact that some of the data points are more informative than others. For example, in interpolating topographic surfaces, the data points taken from the peaks, pits, passes, and pales are more significant than the points taken from the slope or plain. Trend surface is a smoothing and approximate method, it rarely pass through original data points.

It is implicit in multiple regression that the residuals from a regression line or surface are normally distributed and independent errors. The deviations from a trend surface are always to some degree spatially dependent; in fact one of the most fruitful uses of trend surface has been to reveal parts of a study area that show the greatest deviations from the general trend. Therefore, the main use of trend surface is not as an interpolator, but as a way of removing broad features of the data prior to using some other local interpolation methods.

2. *Inverse Distance Weighted:* In this interpolation method, observational points are weighted during interpolation such that the influence of one point relative to another declines with distance from the new point. Weighting is assigned to observational points through the use of a weighting power that controls how weighting factors drop off as the distance from new point increases. The greater the weighting power, the less effect points far from the new point have during interpolation. As the power increases, the value of new point approaches the value of the nearest observational point.

   The simplicity of underlying principle, the speed in calculation, the ease of programming, and reasonable results for many type of data are some of the advantages associated with inverse distance weighted interpolation. The disadvantages are: choice of weighting function may introduce ambiguity, especially when the characteristics of underlying surface are not known; the interpolation can easily affected by uneven distribution of observational data points since an equal weight will be assigned to each of the data points even if it is in a cluster; maxima and minima in the interpolated surface can only occur at data points since inverse distance weighted interpolation is a smoothing technique by definition;

3. *Triangulation:* Triangulation is the most flexible interpolation method offered by ArcInfo. It can generate interpolated surfaces from many different data sources such as point data, lines, breaklines, and polygons (erase, replace, or clip). Because of this flexibility and the speed of interpolation, triangulation has become a popular interpolation method for ArcInfo users.

There are several triangulation methods available (Watson, 1992, p.130-136). Delaunay triangulation is the method adopted in ArcInfo because of these advantages: triangles are as equi-angular as possible, value for a new node is ensured to be as close as possible to a known observation point, and triangulation is not affected by the order of observational points to be considered.

Comparing with grid based interpolation methods, triangulation offers several advantages: First, when the surfaces have significant relief features, triangulation generates more accurate surfaces. This is because the triangulation maintains breakline features such as stream channel, ridges, and shorelines, and relief features such as peaks and pits; Secondly, triangulation is an exact method. It is more accurate than the grid based methods because original data points are located exactly on the surface. The grid based methods only occasionally honour the original data points; Finally, triangulation is also a faster method in terms of computing time. This is because triangulation can efficiently represent the same surface as grid based methods do but with fewer data points.

The main disadvantage is that the surfaces are not smooth and may give jagged appearance. This is caused by discontinuous slopes at the triangle edges and data points. In addition, triangulation is generally not suitable for extrapolation beyond the domain of observed data points.

4. *Kriging:* Kriging is a weighted moving averaging method of interpolation. It is derived from regionlized variable theory which assume that the spatial variation of any geological, soil, or hydrological property, known as a 'regionalized variable' is statistically homogenous throughout the surface,; that is, the same pattern of variation can be observed at all locations on the surface. This spatial variation of the property can be expressed in terms of semivariograms. Kriging derives weights from the semivariograms, and it minimizes the estimation variance which are themselves estimated. In ArcInfo, the estimation variance (called variance grid) is automatically generated as a by-product of interpolated surface.

The computation and interpretation of semivariogram is the "heart" of the kriging method. Semivariogram measures the degree of spatial correlation among observational data points in a study area as a function of the distance and direction between observational data points. It controls the way that kriging weights are assigned to data points during interpolation, and consequently controls the quality of the results.

Computing semivariogram is always the first step for using kriging interpolation methods. Since there are only limited observational data points available, it is impossible to compute the "true" semivariogram. In practice, semivariograms are often estimated from observed data values. This is achieved though selecting the function which fits observational data points the best. This process involves interpretation and judgement, and often requires a large number of "trial and error" computations. ArcInfo offers five mathematical functions which can be used as possible candidates for the semivariogram of a give data set: spherical, circular, exponential, Gaussian, and linear.

Kriging overcomes many shortcomings of the traditional interpolation methods. For instances, the kriging weights are determined by the semivariogram and the configuration of the data set. Kriging is an optimal interpolator in the sense that the estimates are unbiased and have known minimum variances. Since the estimation variances can be determined and mapped like the estimates, and assuming a particular distribution, we can calculate the confidence we can place in the estimates. This makes kriging uniquely different from other interpolation methods. The estimation variance can also be used to determine where more information is needed if future sampling is planned.

Kriging weights depend not only on the distances between observational points and estimation locations but on the mutual distances among observational points as well. As a result, kriging has two interesting and unique properties: declustering and screen effect. With the declustering property, several observational points close together will have collectively the weight of a single observational point located near the centroid of the cluster. With screen effect, the influence of an observational point will be reduced by addition of one or more observational points at the intermediate locations between the original observational point and the estimation location. As the screen effect makes the influence of distant observational points negligible, the use of sampling subset in kriging is a safe practice compared with other weighting methods.

One weakness of kriging is that the original data points are seldom honoured. This is a common problem associated with grid based interpolation methods. It may lead to the dilemma that contours generated from the surface appear on the wrong side of observed data points. Explanation for this is that kriging is essentially a smooth interpolation method, and it behaves like a filter away from the data points. Kriging neither intends nor is able to duplicates reality. Its aim is avoid blunders by following more general spatial trends in the observational data points. Another problem associated with kriging is the estimation of semivariogram. It is not always easy to ascertain whether a particular estimate of the semivariogram is in fact a true estimator of the spatial correlation in an area. The reasons for choosing a particular semivariogram to fit the given data set are often difficult to explain in terms of physical processes. They can only be rationalized in terms of a least-squares on maximum likelihood fit to the data set. Finally, kriging is not a suitable method for data sets which have anomalous pits or spikes, or abrupt changes such as breaklines.

Evaluation of Interpolated Surfaces

If a GIS user can not decide which of above interpolation method should be used , a simple solution is to try several interpolation methods and compare the results through statistical methods. Both univariate and bivariate descriptive tools can be used for such purpose.

1. *Univariate Distribution of Estimation:* It is reasonable to expect that a good interpolation method will produce estimated values whose distribution is similar to the distribution of true values. The differences between estimated values and the true values are referred as residuals. The mean of the residual distribution is often called bias. A reasonable goal for any interpolation method is to produce unbiased estimates. This would mean that the mean, median, and mode of the residual distribution should close to zero. Another feature is to see the small spread of residual distribution. The variance or standard deviation are both good yardsticks for measuring the spread of residual distribution.
2. *Bivariate Distribution of Estimation:* A scatterplot of true verses interpolated values provides a good measure on how well an interpolation method has performed. If all estimated value are the same as true values, the scatterplot would be a 45-degree straight line. In reality, the closer the cloud of points to this line, the better the estimation. In addition, the correlation coefficient can also be used as a reliable index for measuring how close the interpolated values are to the true values.

Chapter 4

Spatial Data

Spatial data, also known as geospatial data, is information about a physical object that can be represented by numerical values in a geographic coordinate system.

Generally speaking, spatial data represents the location, size and shape of an object on planet Earth such as a building, lake, mountain or township. Spatial data may also include attributes that provide more information about the entity that is being represented. Geographic Information Systems (GIS) or other specialised software applications can be used to access, visualize, manipulate and analyze geospatial data.

Spatial Data Types Overview

There are two types of spatial data. The geometry data type supports planar, or Euclidean (flat-earth), data. The geometry data type both conforms to the Open Geospatial Consortium (OGC) Simple Features for SQL Specification version 1.1.0 and is compliant with SQL MM (ISO standard).

In addition, SQL Server supports the geography data type, which stores ellipsoidal (round-earth) data, such as GPS latitude and longitude coordinates.

Spatial Data Objects

The geometry and geography data types support sixteen spatial data objects, or instance types. However, only eleven of these instance types are instantiable; you can create and work with these instances (or instantiate them) in a database. These instances derive certain properties from their parent data types that distinguish them as Points,

LineStrings, CircularStrings, CompoundCurves, Polygons, CurvePolygons or as multiple geometry or geography instances in a GeometryCollection. Geography type has an additional instance type, FullGlobe.

The figure below depicts the geometry hierarchy upon which the geometry and geography data types are based. The instantiable types of geometry and geography are indicated in blue.

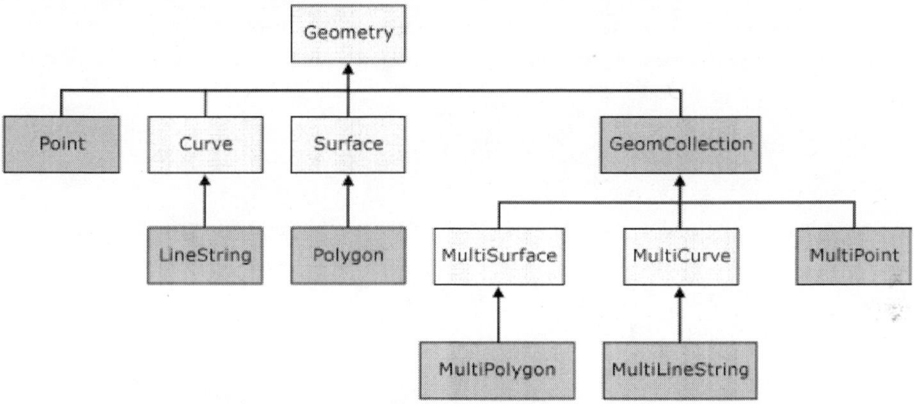

Figure: *As the figure indicates, the ten instantiable types of the geometry and geography data types are Point, MultiPoint, LineString, CircularString, MultiLineString, CompoundCurve, Polygon, CurvePolygon, MultiPolygon, and GeometryCollection. There is one additional instantiable type for the geography data type: FullGlobe. The geometry and geography types can recognise a specific instance as long as it is a well-formed instance, even if the instance is not defined explicitly. For example, if you define a Point instance explicitly using the STPointFromText() method, geometry and geography recognise the instance as a Point, as long as the method input is well-formed. If you define the same instance using the STGeomFromText() method, both the geometry and geography data types recognise the instance as a Point.*

The subtypes for geometry and geography types are divided into simple and collection types. Some methods like STNumCurves() work only with simple types.

Simple types include:

- Point
- LineString
- CircularString
- CompoundCurve
- Polygon
- CurvePolygon

Collection types include:

- MultiPoint
- MultiLineString
- MultiPolygon
- GeometryCollection

Differences between the Geometry and Geography Data Types

The two types of spatial data often behave quite similarly, but there are some key differences in how the data is stored and manipulated.

How Connecting Edges are Defined

The defining data for LineString and Polygon types are vertices only. The connecting edge between two vertices in a geometry type is a straight line. However, the connecting edge between two vertices in a geography type is a short great elliptic arc between the two vertices. A great ellipse is the intersection of the ellipsoid with a plane through its centre and a great elliptic arc is an arc segment on the great ellipse.

How Circular Arc Segments are Defined

Circular arc segments for geometry types are defined on the XY Cartesian coordinate plane (Z values are ignored). Circular arc segments for geography types are defined by curve segments on a reference sphere. Any parallel on the reference sphere can be defined by two complementary circular arcs where the points for both arcs have a constant latitude angle.

Measurements in Spatial Data Types

In the planar, or flat-earth, system, measurements of distances and areas are given in the same unit of measurement as coordinates. Using the geometry data type, the distance between (2, 2) and (5, 6) is 5 units, regardless of the units used.

In the ellipsoidal, or round-earth system, coordinates are given in degrees of latitude and longitude. However, lengths and areas are usually measured in metres and square metres, though the measurement may depend on the spatial reference identifier (SRID) of the geography instance. The most common unit of measurement for the geography data type is metres.

Orientation of Spatial Data

In the planar system, the ring orientation of a polygon is not an important factor. For example, a polygon described by ((0, 0), (10, 0),

(0, 20), (0, 0)) is the same as a polygon described by ((0, 0), (0, 20), (10, 0), (0, 0)). The OGC Simple Features for SQL Specification does not dictate a ring ordering, and SQL Server does not enforce ring ordering.

In an ellipsoidal system, a polygon has no meaning, or is ambiguous, without an orientation. For example, does a ring around the equator describe the northern or southern hemisphere? If we use the geography data type to store the spatial instance, we must specify the orientation of the ring and accurately describe the location of the instance. The interior of the polygon in an ellipsoidal system is defined by the left-hand rule.

When the compatibility level is 100 or below in SQL Server 2014 then the geography data type has the following restrictions:

- Each geography instance must fit inside a single hemisphere. No spatial objects larger than a hemisphere can be stored.
- Any geography instance from an Open Geospatial Consortium (OGC) Well-Known Text (WKT) or Well-Known Binary (WKB) representation that produces an object larger than a hemisphere throws an ArgumentException.
- The geography data type methods that require the input of two geography instances, such as STIntersection(), STUnion(), STDifference(), and STSymDifference(), will return null if the results from the methods do not fit inside a single hemisphere. STBuffer() will also return null if the output exceeds a single hemisphere.

In SQL Server 2014, FullGlobe is a special type of Polygon that covers the entire globe. FullGlobe has an area, but no borders or vertices.

Outer and Inner Rings not Important in Geography Data Type

The OGC Simple Features for SQL Specification discusses outer rings and inner rings, but this distinction makes little sense for the SQL Server geography data type; any ring of a polygon can be taken to be the outer ring.

Circular Arc Segments

Three instantiable types can take circular arc segments: CircularString, CompoundCurve, and CurvePolygon. A circular arc segment is defined by three points in a two dimensional plane and the third point cannot be the same as the first point.

Methods operating on circular arc segment types use straight line segments to approximate the circular arc. The number of line segments used to approximate the arc will depend on the length and curvature of the arc. Z values can be stored for each of the circular arc segment types; however, methods will not use the Z values in their calculations.

If Z values are given for circular arc segments then they must be the same for all points in the circular arc segment for it to be accepted for input. For example: CIRCULARSTRING(0 0 1, 2 2 1, 4 0 1) is accepted, but CIRCULARSTRING(0 0 1, 2 2 2, 4 0 1) is not accepted.

LineString and CircularString Comparison

The following diagram shows identical isosceles triangles (triangle A uses line segments to define the triangle and triangle B uses circular arc segments to defined the triangle):

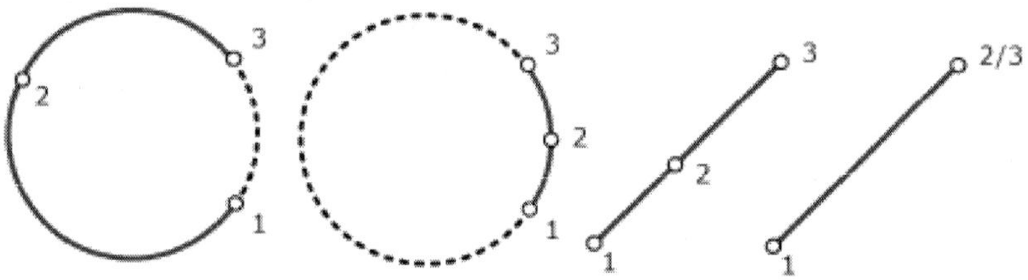

This example shows how to store the above isosceles triangles using both a LineString instance and CircularString instance:

Transact-SQL

```
DECLARE @g1 geometry;
DECLARE @g2 geometry;
SET @g1 = geometry::STGeomFromText('LINESTRING(1 1, 5 1, 3 5, 1 1)', 0);
SET @g2 = geometry::STGeomFromText('CIRCULARSTRING(1 1, 3 1, 5 1, 4 3, 3 5, 2
3, 1 1)', 0);
IF @g1.STIsValid() = 1 AND @g2.STIsValid() = 1
  BEGIN
      SELECT @g1.ToString(), @g2.ToString()
      SELECT @g1.STLength() AS [LS Length], @g2.STLength() AS [CS Length]
  END
```

Notice that a CircularString instance requires seven points to define the triangle, but a LineString instance requires only four points to define the triangle. The reason for this is that a CircularString instance stores circular arc segments and not line segments. So the sides of the triangle stored in the CircularString instance are ABC, CDE, and EFA whereas the sides of the triangle stored in the LineString instance are AC, CE, and EA.

Consider the following code snippet:

```
SET @g1 = geometry::STGeomFromText('LINESTRING(0 0, 2 2, 4 0)', 0);
SET @g2 = geometry::STGeomFromText('CIRCULARSTRING(0 0, 2 2, 4 0)', 0);
SELECT @g1.STLength() AS [LS Length], @g2.STLength() AS [CS Length];
```

This snippet will produce the following results:

LS LengthCS Length

5.65685…6.28318…

The following illustration shows how each type is stored (red line shows LineString @g1, blue line shows CircularString @g2):

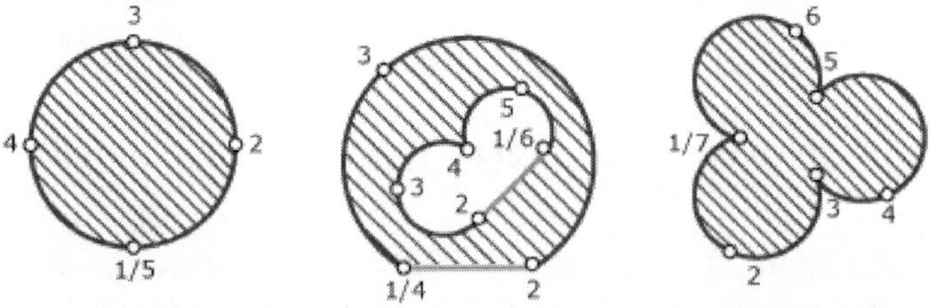

As the illustration above shows, CircularString instances use fewer points to store curve boundaries with greater precision than LineString instances. CircularString instances are useful for storing circular boundaries like a twenty-mile search radius from a specific point. LineString instances are good for storing boundaries that are linear like a square city block.

LineString and CompoundCurve Comparison

The following code examples show how to store the same figure using LineString and CompoundCurve instances:

Transact-SQL:

```
SET @g = geometry::Parse('LINESTRING(2 2, 4 2, 4 4, 2 4, 2 2)');
SET @g = geometry::Parse('COMPOUNDCURVE((2 2, 4 2), (4 2, 4 4), (4 4, 2 4),
(2 4, 2 2))');
SET @g = geometry::Parse('COMPOUNDCURVE((2 2, 4 2, 4 4, 2 4, 2 2))');
```

or

In the examples above, either a LineString instance or a CompoundCurve instance could store the figure. This next example uses a CompoundCurve to store a pie slice:

Transact-SQL

```
SET @g = geometry::Parse('COMPOUNDCURVE(CIRCULARSTRING(2 2, 1 3, 0 2),(0 2, 1
0, 2 2))');
```

A CompoundCurve instance can store the circular arc segment (2 2, 1 3, 0 2) directly whereas a LineString instance would have to convert the curve into several smaller line segments.

CircularString and CompoundCurve Comparison

The following code example shows how the pie slice can be stored in a CircularString instance:

Transact-SQL:

```
DECLARE @g geometry;
SET @g = geometry::Parse('CIRCULARSTRING( 0 0, 1 2.1082, 3 6.3246, 0 7, -3
6.3246, -1 2.1082, 0 0)');
SELECT @g.ToString(), @g.STLength();
```

To store the pie slice using a CircularString instance requires that three points be used for each line segment. If an intermediate point is not known, it either has to be calculated or the endpoint of the line segment has to be doubled as the following snippet shows:

Transact-SQL:

```
SET @g = geometry::Parse('CIRCULARSTRING( 0 0, 3 6.3246, 3 6.3246, 0 7, -3
6.3246, 0 0, 0 0)');
```

CompoundCurve instances allow both LineString and CircularString components so that only two points to the line segments of the pie slice need to be known. This code example shows how to use a CompoundCurve to store the same Figure:

Transact-SQL

```
DECLARE @g geometry;
SET @g = geometry::Parse('COMPOUNDCURVE(CIRCULARSTRING( 3 6.3246, 0 7, -3
6.3246), (-3 6.3246, 0 0, 3 6.3246))');
SELECT @g.ToString(), @g.STLength();
```

Polygon and CurvePolygon Comparison

CurvePolygon instances can use CircularString and CompoundCurve instances when defining their exterior and interior rings. Polygon instances cannot use the circular arc segment types: CircularString and CompoundCurve.

The Role of Geographic Information Systems

A geographic information system (GIS) is a computer system designed to capture, store, manipulate, analyze, manage, and present all types of spatial or geographical data. The acronym GIS is sometimes used for geographical information science or geospatial information studies to refer to the academic discipline or career of working with geographic information systems and is a large domain within the broader academic discipline of Geoinformatics. What goes beyond a GIS is a spatial data infrastructure, a concept that has no such restrictive boundaries.

In a general sense, the term describes any information system that integrates, stores, edits, analyzes, shares, and displays geographic information. GIS applications are tools that allow users to create interactive queries (user-created searches), analyze spatial information, edit data in maps, and present the results of all these operations. Geographic information science is the science underlying geographic concepts, applications, and systems.

GIS is a broad term that can refer to a number of different technologies, processes, and methods. It is attached to many operations and has many applications related to engineering, planning, management, transport/logistics, insurance, telecommunications, and business. For that reason, GIS and location intelligence applications can be the foundation for many location-enabled services that rely on analysis and visualization.

GIS can relate unrelated information by using location as the key index variable. Locations or extents in the Earth space–time may be recorded as dates/times of occurrence, and x, y, and z coordinates representing, longitude, latitude, and elevation, respectively. All Earth-based spatial–temporal location and extent references should, ideally, be relatable to one another and ultimately to a "real" physical location or extent. This key characteristic of GIS has begun to open new avenues of scientific inquiry.

History of Development

The first known use of the term "geographic information system" was by Roger Tomlinson in the year 1968 in his paper "A Geographic Information System for Regional Planning". Tomlinson is also acknowledged as the "father of GIS".

***Figure:** E. W. Gilbert's version (1958) of John Snow's 1855 map of the Soho cholera outbreak showing the clusters of cholera cases in the London epidemic of 1854*

Previously, one of the first applications of spatial analysis in epidemiology is the 1832 "*Rapport sur la marche et les effets du choléra dans Paris et le département de la Seine*". The French geographer Charles Picquet represented the 48 districts of the city of Paris by halftone colour gradient according to the percentage of deaths by cholera per 1,000 inhabitants. In 1854 John Snow depicted a cholera outbreak in London using points to represent the locations of some individual cases, an early successful use of a geographic methodology in epidemiology. While the basic elements of topography and theme existed previously in cartography, the John Snow map was unique, using cartographic methods not only to depict but also to analyze clusters of geographically dependent phenomena.

The early 20th century saw the development of photozincography, which allowed maps to be split into layers, for example one layer for vegetation and another for water. This was particularly used for printing contours – drawing these was a labour-intensive task but having them on a separate layer meant they could be worked on without the other layers to confuse the draughtsman. This work was originally drawn on glass plates but later plastic film was introduced, with the advantages of being lighter, using less storage space and being less brittle, among others. When all the layers were finished, they were combined into one image using a large process camera. Once colour printing came in, the layers idea was also used for creating separate printing plates for each colour. While the use of layers much later became one of the main typical features of a contemporary GIS, the photographic process just described is not considered to be a GIS in itself – as the maps were just images with no database to link them to.

Computer hardware development spurred by nuclear weapon research led to general-purpose computer "mapping" applications by the early 1960s.

The year 1960 saw the development of the world's first true operational GIS in Ottawa, Ontario, Canada by the federal Department of Forestry and Rural Development. Developed by Dr. Roger Tomlinson, it was called the Canada Geographic Information System (CGIS) and was used to store, analyze, and manipulate data collected for the Canada Land Inventory – an effort to determine the land capability for rural Canada by mapping information about soils, agriculture, recreation, wildlife, waterfowl, forestry and land use at a scale of 1:50,000. A rating classification factor was also added to permit analysis.

CGIS was an improvement over "computer mapping" applications as it provided capabilities for overlay, measurement, and digitizing/

scanning. It supported a national coordinate system that spanned the continent, coded lines as arcs having a true embedded topology and it stored the attribute and locational information in separate files. As a result of this, Tomlinson has become known as the "father of GIS", particularly for his use of overlays in promoting the spatial analysis of convergent geographic data.

CGIS lasted into the 1990s and built a large digital land resource database in Canada. It was developed as a mainframe-based system in support of federal and provincial resource planning and management. Its strength was continent-wide analysis of complex datasets. The CGIS was never available commercially.

In 1964 Howard T. Fisher formed the Laboratory for Computer Graphics and Spatial Analysis at the Harvard Graduate School of Design (LCGSA 1965–1991), where a number of important theoretical concepts in spatial data handling were developed, and which by the 1970s had distributed seminal software code and systems, such as SYMAP, GRID, and ODYSSEY – that served as sources for subsequent commercial development—to universities, research centres and corporations worldwide.

By the early 1980s, M&S Computing (later Intergraph) along with Bentley Systems Incorporated for the CAD platform, Environmental Systems Research Institute (ESRI), CARIS (Computer Aided Resource Information System), MapInfo Corporation and ERDAS (Earth Resource Data Analysis System) emerged as commercial vendors of GIS software, successfully incorporating many of the CGIS features, combining the first generation approach to separation of spatial and attribute information with a second generation approach to organising attribute data into database structures. In parallel, the development of two public domain systems (MOSS and GRASS GIS) began in the late 1970s and early 1980s.

In 1986, Mapping Display and Analysis System (MIDAS), the first desktop GIS product emerged for the DOS operating system. This was renamed in 1990 to MapInfo for Windows when it was ported to the Microsoft Windows platform. This began the process of moving GIS from the research department into the business environment.

By the end of the 20th century, the rapid growth in various systems had been consolidated and standardized on relatively few platforms and users were beginning to explore viewing GIS data over the Internet, requiring data format and transfer standards. More recently, a growing number of free, open-source GIS packages run on a range of operating systems and can be customized to perform specific tasks. Increasingly geospatial data and mapping applications are being made available via the world wide web.

Several authoritative articles on the history of GIS have been published.

GIS Techniques and Technology

Modern GIS technologies use digital information, for which various digitized data creation methods are used. The most common method of data creation is digitization, where a hard copy map or survey plan is transferred into a digital medium through the use of a CAD program, and geo-referencing capabilities. With the wide availability of ortho-rectified imagery (both from satellite and aerial sources), heads-up digitizing is becoming the main avenue through which geographic data is extracted. Heads-up digitizing involves the tracing of geographic data directly on top of the aerial imagery instead of by the traditional method of tracing the geographic form on a separate digitizing tablet (heads-down digitizing).

Relating Information from Different Sources

GIS uses spatio-temporal (space-time) location as the key index variable for all other information. Just as a relational database containing text or numbers can relate many different tables using common key index variables, GIS can relate otherwise unrelated information by using location as the key index variable. The key is the location and/or extent in space-time.

Any variable that can be located spatially, and increasingly also temporally, can be referenced using a GIS. Locations or extents in Earth space–time may be recorded as dates/times of occurrence, and x, y, and z coordinates representing, longitude, latitude, and elevation, respectively. These GIS coordinates may represent other quantified systems of temporo-spatial reference (for example, film frame number, stream gage station, highway mile-marker, surveyor benchmark, building address, street intersection, entrance gate, water depth sounding, POS or CAD drawing origin/units). Units applied to recorded temporal-spatial data can vary widely (even when using exactly the same data), but all Earth-based spatial–temporal location and extent references should, ideally, be relatable to one another and ultimately to a "real" physical location or extent in space–time.

Related by accurate spatial information, an incredible variety of real-world and projected past or future data can be analyzed, interpreted and represented. This key characteristic of GIS has begun to open new avenues of scientific inquiry into behaviours and patterns of real-world information that previously had not been systematically correlated.

GIS Uncertainties

GIS accuracy depends upon source data, and how it is encoded to be data referenced. Land surveyors have been able to provide a high level of positional accuracy utilizing the GPS-derived positions. High-resolution digital terrain and aerial imagery, powerful computers and Web technology are changing the quality, utility, and expectations of GIS to serve society on a grand scale, but nevertheless there are other source data that have an impact on overall GIS accuracy like paper maps, though these may be of limited use in achieving the desired accuracy since the aging of maps affects their dimensional stability.

In developing a digital topographic database for a GIS, topographical maps are the main source, and aerial photography and satellite imagery are extra sources for collecting data and identifying attributes which can be mapped in layers over a location facsimile of scale. The scale of a map and geographical rendering area representation type are very important aspects since the information content depends mainly on the scale set and resulting locatability of the map's representations. In order to digitize a map, the map has to be checked within theoretical dimensions, then scanned into a raster format, and resulting raster data has to be given a theoretical dimension by a rubber sheeting/warping technology process.

A quantitative analysis of maps brings accuracy issues into focus. The electronic and other equipment used to make measurements for GIS is far more precise than the machines of conventional map analysis. All geographical data are inherently inaccurate, and these inaccuracies will propagate through GIS operations in ways that are difficult to predict.

Data Representation

GIS data represents real objects (such as roads, land use, elevation, trees, waterways, etc.) with digital data determining the mix. Real objects can be divided into two abstractions: discrete objects (e.g., a house) and continuous fields (such as rainfall amount, or elevations). Traditionally, there are two broad methods used to store data in a GIS for both kinds of abstractions mapping references: raster images and vector. Points, lines, and polygons are the stuff of mapped location attribute references. A new hybrid method of storing data is that of identifying point clouds, which combine three-dimensional points with

RGB information at each point, returning a "3D colour image". GIS thematic maps then are becoming more and more realistically visually descriptive of what they set out to show or determine.

Data Capture

Data capture—entering information into the system—consumes much of the time of GIS practitioners. There are a variety of methods used to enter data into a GIS where it is stored in a digital format.

Existing data printed on paper or PET film maps can be digitized or scanned to produce digital data. A digitizer produces vector data as an operator traces points, lines, and polygon boundaries from a map. Scanning a map results in raster data that could be further processed to produce vector data.

Survey data can be directly entered into a GIS from digital data collection systems on survey instruments using a technique called coordinate geometry (COGO). Positions from a global navigation satellite system (GNSS) like Global Positioning System can also be collected and then imported into a GIS. A current trend in data collection gives users the ability to utilize field computers with the ability to edit live data using wireless connections or disconnected editing sessions. This has been enhanced by the availability of low-cost mapping-grade GPS units with decimeter accuracy in real time. This eliminates the need to post process, import, and update the data in the office after fieldwork has been collected. This includes the ability to incorporate positions collected using a laser rangefinder. New technologies also allow users to create maps as well as analysis directly in the field, making projects more efficient and mapping more accurate.

Remotely sensed data also plays an important role in data collection and consist of sensors attached to a platform. Sensors include cameras, digital scanners and lidar, while platforms usually consist of aircraft and satellites. Recently with the development of miniature UAVs, aerial data collection is becoming possible at much lower costs, and on a more frequent basis. For example, the Aeryon Scout was used to map a 50-acre area with a Ground sample distance of 1 inch (2.54 cm) in only 12 minutes.

The majority of digital data currently comes from photo interpretation of aerial photographs. Soft-copy workstations are used to digitize features directly from stereo pairs of digital photographs. These systems allow data to be captured in two and three dimensions, with elevations measured directly from a stereo pair using principles of photogrammetry. Analog aerial photos must be scanned before being entered into a soft-copy system, for high-quality digital cameras this step is skipped.

Satellite remote sensing provides another important source of spatial data. Here satellites use different sensor packages to passively measure the reflectance from parts of the electromagnetic spectrum or radio waves that were sent out from an active sensor such as radar. Remote sensing collects raster data that can be further processed using different bands to identify objects and classes of interest, such as land cover. When data is captured, the user should consider if the data should be captured with either a relative accuracy or absolute accuracy, since this could not only influence how information will be interpreted but also the cost of data capture.

After entering data into a GIS, the data usually requires editing, to remove errors, or further processing. For vector data it must be made "topologically correct" before it can be used for some advanced analysis. For example, in a road network, lines must connect with nodes at an intersection. Errors such as undershoots and overshoots must also be removed. For scanned maps, blemishes on the source map may need to be removed from the resulting raster. For example, a fleck of dirt might connect two lines that should not be connected.

Raster-to-Vector Translation

Data restructuring can be performed by a GIS to convert data into different formats. For example, a GIS may be used to convert a satellite image map to a vector structure by generating lines around all cells with the same classification, while determining the cell spatial relationships, such as adjacency or inclusion.

More advanced data processing can occur with image processing, a technique developed in the late 1960s by NASA and the private sector to provide contrast enhancement, false colour rendering and a variety of other techniques including use of two dimensional Fourier transforms. Since digital data is collected and stored in various ways, the two data sources may not be entirely compatible. So a GIS must be able to convert geographic data from one structure to another. In so doing, the implicit assumptions behind different ontologies and classifications require analysis. Object ontologies have gained increasing prominence as a consequence of object-oriented programming and sustained work by Barry Smith and co-workers.

Projections, Coordinate Systems and Registration

The earth can be represented by various models, each of which may provide a different set of coordinates (e.g., latitude, longitude, elevation) for any given point on the Earth's surface. The simplest model is to assume the earth is a perfect sphere. As more measurements of the earth have accumulated, the models of the earth have become more sophisticated and more accurate. In fact, there are models called datums that apply to different areas of the earth to provide increased accuracy, like NAD83 for U.S. measurements, and the World Geodetic System for worldwide measurements.

Spatial Analysis with GIS

GIS spatial analysis is a rapidly changing field, and GIS packages are increasingly including analytical tools as standard built-in facilities, as optional toolsets, as add-ins or 'analysts'. In many instances these are provided by the original software suppliers (commercial vendors or collaborative non commercial development teams), whilst in other cases facilities have been developed and are provided by third parties. Furthermore, many products offer software development kits (SDKs), programming languages and language support, scripting facilities and/or special interfaces for developing one's own analytical tools or variants. The website "Geospatial Analysis" and associated book/ebook attempt to provide a reasonably comprehensive guide to the subject. The increased availability has created a new dimension to business intelligence termed "spatial intelligence" which, when openly delivered via intranet, democratizes access to geographic and social network data. Geospatial intelligence, based on GIS spatial analysis, has also become a key element for security. GIS as a whole can be described as conversion to a vectorial representation or to any other digitisation process.

Slope and Aspect

Slope can be defined as the steepness or gradient of a unit of terrain, usually measured as an angle in degrees or as a percentage. Aspect can be defined as the direction in which a unit of terrain faces. Aspect is usually expressed in degrees from north. Slope, aspect, and surface curvature in terrain analysis are all derived from neighbourhood operations using elevation values of a cell's adjacent neighbours. Slope is a function of resolution, and the spatial resolution used to calculate slope and aspect should always be specified. Authors such as Skidmore, Jones and Zhou and Liu have compared techniques for calculating slope and aspect.

The Following Method can be Used to Derive Slope and Aspect:

The elevation at a point or unit of terrain will have perpendicular tangents (slope) passing through the point, in an east-west and north-south direction. These two tangents give two components,

$\partial z / \partial x$ and $\partial z / \partial y$, which then be used to determine the overall direction of slope, and the aspect of the slope. The gradient is defined as a vector quantity with components equal to the partial derivatives of the surface in the x and y directions.

The calculation of the overall 3x3 grid slope S and aspect A for methods that determine east-west and north-south component use the following formulas respectively:

$$\tan S = \sqrt{\left(\frac{\partial z}{\partial x}\right)^2 + \left(\frac{\partial z}{\partial y}\right)^2}$$

$$\tan A = \left(\frac{\left(\frac{-\partial z}{\partial y}\right)}{\left(\frac{\partial z}{\partial x}\right)}\right)$$

Zhou and Liu describe another algorithm for calculating aspect, as follows:

$$A = 270^{\circ} + \arctan\left(\frac{\left(\frac{\partial z}{\partial x}\right)}{\left(\frac{\partial z}{\partial y}\right)}\right) - 90^{\circ}\left(\frac{\left(\frac{\partial z}{\partial y}\right)}{\left|\frac{\partial z}{\partial y}\right|}\right)$$

Data Analysis

It is difficult to relate wetlands maps to rainfall amounts recorded at different points such as airports, television stations, and schools. A GIS, however, can be used to depict two- and three-dimensional characteristics of the Earth's surface, subsurface, and atmosphere from information points. For example, a GIS can quickly generate a map with isopleth or contour lines that indicate differing amounts of rainfall. Such a map can be thought of as a rainfall contour map. Many sophisticated methods can estimate the characteristics of surfaces from a limited number of point measurements. A two-dimensional contour map created from the surface modelling of rainfall point measurements may be overlaid and analyzed with any other map in a GIS covering the same area. This GIS derived map can then provide additional information - such as the viability of water power potential as a renewable energy source. Similarly, GIS can be used compare other renewable energy resources to find the best geographic potential for a region. Additionally, from a series of three-dimensional points, or digital elevation model, isopleth lines representing elevation contours can be generated, along with slope analysis, shaded relief, and other elevation products. Watersheds can be easily defined for any given reach, by computing all of the areas contiguous and uphill from any given point of interest. Similarly, an expected thalweg of where surface water would want to travel in intermittent and permanent streams can be computed from elevation data in the GIS.

Topological Modelling

A GIS can recognise and analyze the spatial relationships that exist within digitally stored spatial data. These topological relationships allow complex spatial modelling and analysis to be performed. Topological relationships between geometric entities traditionally include adjacency (what adjoins what), containment (what encloses what), and proximity (how close something is to something else).

Geometric Networks

Geometric networks are linear networks of objects that can be used to represent interconnected features, and to perform special spatial analysis on them. A geometric network is composed of edges, which are connected at junction points, similar to graphs in mathematics and computer science. Just like graphs, networks can have weight and flow assigned to its edges, which can be used to represent various interconnected features more accurately. Geometric networks are often used to model road networks and public utility networks, such as electric, gas, and water networks. Network modelling is also commonly employed in transportation planning, hydrology modelling, and infrastructure modelling.

Hydrological Modelling

GIS hydrological models can provide a spatial element that other hydrological models lack, with the analysis of variables such as slope, aspect and watershed or catchment area. Terrain analysis is fundamental to hydrology, since water always flows down a slope. As basic terrain analysis of a digital elevation model (DEM) involves calculation of slope and aspect, DEMs are very useful for hydrological analysis. Slope and aspect can then be used to determine direction of surface runoff, and hence flow accumulation for the formation of streams, rivers and lakes. Areas of divergent flow can also give a clear indication of the boundaries of a catchment. Once a flow direction and accumulation matrix has been created, queries can be performed that show contributing or dispersal areas at a certain point. More detail can be added to the model, such as terrain roughness, vegetation types and soil types, which can influence infiltration and evapotranspiration rates, and hence influencing surface flow. One of the main uses of hydrological modelling is in environmental contamination research.

Cartographic Modelling

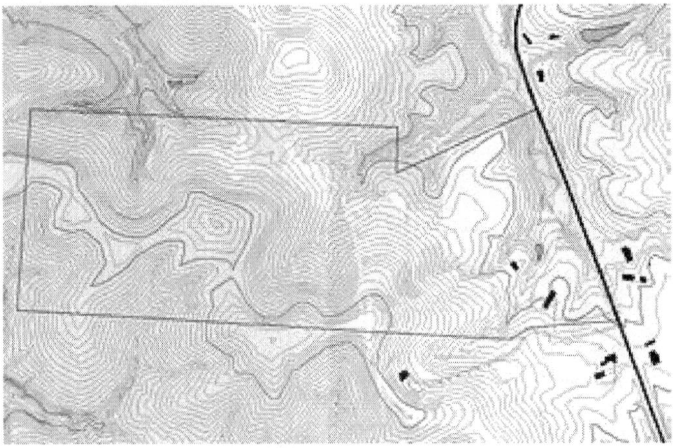

Figure: *An example of use of layers in a GIS application. In this example, the forest cover layer (light green) is at the bottom, with the topographic layer over it. Next up is the stream layer, then the boundary layer, then the road layer. The order is very important in order to properly display the final result. Note that the pond layer was located just below the stream layer, so that a stream line can be seen overlying one of the ponds.*

The term "cartographic modelling" was probably coined by Dana Tomlin in his PhD dissertation and later in his book which has the term in the title. Cartographic modelling refers to a process where several thematic layers of the same area are produced, processed, and analyzed. Tomlin used raster layers, but the overlay method can be used more generally. Operations on map layers can be combined into algorithms, and eventually into simulation or optimization models.

Map Overlay

The combination of several spatial datasets (points, lines, or polygons) creates a new output vector dataset, visually similar to stacking several maps of the same region. These overlays are similar to mathematical Venn diagram overlays. A union overlay combines the geographic features and attribute tables of both inputs into a single new output. An intersect overlay defines the area where both inputs overlap and retains a set of attribute fields for each. A symmetric difference overlay defines an output area that includes the total area of both inputs except for the overlapping area.

Data extraction is a GIS process similar to vector overlay, though it can be used in either vector or raster data analysis. Rather than combining the properties and features of both datasets, data extraction involves using a "clip" or "mask" to extract the features of one data set that fall within the spatial extent of another dataset.

In raster data analysis, the overlay of datasets is accomplished through a process known as "local operation on multiple rasters" or "map algebra," through a function that combines the values of each raster's matrix. This function may weigh some inputs more than others through use of an "index model" that reflects the influence of various factors upon a geographic phenomenon.

Geostatistics

Geostatistics is a branch of statistics that deals with field data, spatial data with a continuous index. It provides methods to model spatial correlation, and predict values at arbitrary locations (interpolation).

When phenomena are measured, the observation methods dictate the accuracy of any subsequent analysis. Due to the nature of the data (e.g. traffic patterns in an urban environment; weather patterns over the Pacific Ocean), a constant or dynamic degree of precision is always lost in the measurement. This loss of precision is determined from the scale and distribution of the data collection.

To determine the statistical relevance of the analysis, an average is determined so that points (gradients) outside of any immediate measurement can be included to determine their predicted behaviour. This is due to the limitations of the applied statistic and data collection methods, and interpolation is required to predict the behaviour of particles, points, and locations that are not directly measurable.

Interpolation is the process by which a surface is created, usually a raster dataset, through the input of data collected at a number of sample points. There are several forms of interpolation, each which treats the data differently, depending on the properties of the data set. In comparing interpolation methods, the first consideration should be whether or not the source data will change (exact or approximate). Next is whether the method is subjective, a human interpretation, or objective. Then there is the nature of transitions between points: are they abrupt or gradual. Finally, there is whether a method is global (it uses the entire data set to form the model), or local where an algorithm is repeated for a small section of terrain.

Interpolation is a justified measurement because of a spatial autocorrelation principle that recognises that data collected at any position will have a great similarity to, or influence of those locations within its immediate vicinity. Digital elevation models, triangulated irregular networks, edge-finding algorithms, Thiessen polygons, Fourier analysis, (weighted) moving averages, inverse distance weighting, kriging, spline, and trend surface analysis are all mathematical methods to produce interpolative data.

Address Geocoding

Geocoding is interpolating spatial locations (X,Y coordinates) from street addresses or any other spatially referenced data such as ZIP Codes, parcel lots and address locations. A reference theme is required to geocode individual addresses, such as a road centreline file with address ranges. The individual address locations have historically been interpolated, or estimated, by examining address ranges along a road segment. These are usually provided in the form of a table or database. The software will then place a dot approximately where that address belongs along the segment of centreline. For example, an address point of 500 will be at the midpoint of a line segment that starts with address 1 and ends with address 1,000. Geocoding can also be applied against actual parcel data, typically from municipal tax maps. In this case, the result of the geocoding will be an actually positioned space as opposed to an interpolated point. This approach is being increasingly used to provide more precise location information.

Reverse Geocoding

Reverse geocoding is the process of returning an estimated street address number as it relates to a given coordinate. For example, a user can click on a road centreline theme (thus providing a coordinate) and have information returned that reflects the estimated house number. This house number is interpolated from a range assigned to that road segment. If the user clicks at the midpoint of a segment that starts with address 1 and ends with 100, the returned value will be somewhere near 50. Note that reverse geocoding does not return actual addresses, only estimates of what should be there based on the predetermined range.

Multi-Criteria Decision Analysis

Coupled with GIS, multi-criteria decision analysis methods support decision-makers in analysing a set of alternative spatial solutions, such as the most likely ecological habitat for restoration, against multiple criteria, such as vegetation cover or roads. MCDA uses decision rules to aggregate the criteria, which allows the alternative solutions to be ranked or prioritised. GIS MCDA may reduce costs and time involved in identifying potential restoration sites.

Data Output and Cartography

Cartography is the design and production of maps, or visual representations of spatial data. The vast majority of modern cartography is done with the help of computers, usually using GIS but production of quality cartography is also achieved by importing layers into a design program to refine it. Most GIS software gives the user substantial control over the appearance of the data.

Cartographic work serves two major functions:

First, it produces graphics on the screen or on paper that convey the results of analysis to the people who make decisions about resources. Wall maps and other graphics can be generated, allowing the viewer to visualize and thereby understand the results of analyses or simulations of potential events. Web Map Servers facilitate distribution of generated maps through web browsers using various implementations of web-based application programming interfaces (AJAX, Java, Flash, etc.).

Second, other database information can be generated for further analysis or use. An example would be a list of all addresses within one mile (1.6 km) of a toxic spill.

Graphic Display Techniques

Traditional maps are abstractions of the real world, a sampling of important elements portrayed on a sheet of paper with symbols to represent physical objects. People who use maps must interpret these symbols. Topographic maps show the shape of land surface with contour lines or with shaded relief.

Today, graphic display techniques such as shading based on altitude in a GIS can make relationships among map elements visible, heightening one's ability to extract and analyze information. For example, two types of data were combined in a GIS to produce a perspective view of a portion of San Mateo County, California.

- The digital elevation model, consisting of surface elevations recorded on a 30-metre horizontal grid, shows high elevations as white and low elevation as black.
- The accompanying Landsat Thematic Mapper image shows a false-colour infrared image looking down at the same area in 30-metre pixels, or picture elements, for the same coordinate points, pixel by pixel, as the elevation information.

A GIS was used to register and combine the two images to render the three-dimensional perspective view looking down the San Andreas

Fault, using the Thematic Mapper image pixels, but shaded using the elevation of the landforms. The GIS display depends on the viewing point of the observer and time of day of the display, to properly render the shadows created by the sun's rays at that latitude, longitude, and time of day.

An archeochrome is a new way of displaying spatial data. It is a thematic on a 3D map that is applied to a specific building or a part of a building. It is suited to the visual display of heat-loss data.

Spatial ETL

Spatial ETL tools provide the data processing functionality of traditional Extract, Transform, Load (ETL) software, but with a primary focus on the ability to manage spatial data. They provide GIS users with the ability to translate data between different standards and proprietary formats, whilst geometrically transforming the data en route. These tools can come in the form of add-ins to existing wider-purpose software such as Microsoft Excel.

GIS Data Mining

GIS or spatial data mining is the application of data mining methods to spatial data. Data mining, which is the partially automated search for hidden patterns in large databases, offers great potential benefits for applied GIS-based decision making. Typical applications including environmental monitoring. A characteristic of such applications is that spatial correlation between data measurements require the use of specialised algorithms for more efficient data analysis.

GIS File Formats: A GIS file format is a standard of encoding geographical information into a file. They are created mainly by government mapping agencies (such as the USGS or National Geospatial-Intelligence Agency) or by GIS software developers.

Raster

A raster data type is, in essence, any type of digital image represented by reducible and enlargeable grids. Anyone who is familiar with digital photography will recognise the Raster graphics pixel as the smallest individual grid unit building block of an image, usually not readily identified as an artifact shape until an image is produced on a very large scale. A combination of the pixels making up an image colour formation scheme will compose details of an image, as is distinct from the commonly used points, lines, and polygon area location symbols of scalable vector graphics as the basis of the vector model of area attribute rendering. While a digital image is concerned with its output blending together its grid based details as an identifiable representation of reality, in a photograph or art image transferred into a computer, the raster data type will reflect a digitized abstraction of reality dealt with by grid populating tones or objects, quantities, cojoined or open boundaries, and map relief schemas. Aerial photos are one commonly used form of raster data, with one primary purpose in mind: to display a detailed image on a map area, or for the purposes of rendering its identifiable objects by digitization. Additional raster data sets used by a GIS will contain information regarding elevation, a digital elevation model, or reflectance of a particular wavelength of light, Landsat, or other electromagnetic spectrum indicators.

Raster data type consists of rows and columns of cells, with each cell storing a single value. Raster data can be images (raster images) with each pixel (or cell) containing a colour value. Additional values recorded for each cell may be a discrete value, such as land use, a continuous value, such as temperature, or a null value if no data is available. While a raster cell stores a single value, it can be extended by using raster bands to represent RGB (red, green, blue) colours, colourmaps (a mapping between a thematic code and RGB value), or an extended attribute table with one row for each unique cell value. The resolution of the raster data set is its cell width in ground units.

Raster data is stored in various formats; from a standard file-based structure of TIFF, JPEG, etc. to binary large object (BLOB) data stored directly in a relational database management system (RDBMS) similar to other vector-based feature classes. Database storage, when properly indexed, typically allows for quicker retrieval of the raster data but can require storage of millions of significantly sized records.

Vector

In a GIS, geographical features are often expressed as vectors, by considering those features as geometrical shapes. Different geographical features are expressed by different types of geometry:

Points

Zero-dimensional points are used for geographical features that can best be expressed by a single point reference—in other words, by simple location. Examples include wells, peaks, features of interest, and trailheads. Points convey the least amount of information of these file types. Points can also be used to represent areas when displayed at a small scale. For example, cities on a map of the world might be represented by points rather than polygons. No measurements are possible with point features.

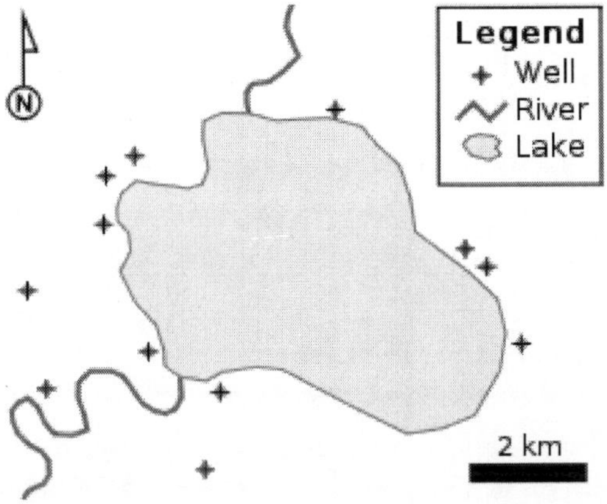

Figure: *A simple vector map, using each of the vector elements: points for wells, lines for rivers, and a polygon for the lake*

Lines or Polylines

One-dimensional lines or polylines are used for linear features such as rivers, roads, railroads, trails, and topographic lines. Again, as with point features, linear features displayed at a small scale will be represented as linear features rather than as a polygon. Line features can measure distance.

Polygons

Two-dimensional polygons are used for geographical features that cover a particular area of the earth's surface. Such features may include lakes, park boundaries, buildings, city boundaries, or land uses. Polygons convey the most amount of information of the file types. Polygon features can measure perimeter and area.

Each of these geometries are linked to a row in a database that describes their attributes. For example, a database that describes lakes may contain a lake's depth, water quality, pollution level. This information can be used to make a map to describe a particular attribute of the dataset. For example, lakes could be coloured depending on level of pollution. Different geometries can also be compared. For example, the GIS could be used to identify all wells (point geometry) that are within one kilometre of a lake (polygon geometry) that has a high level of pollution.

Vector features can be made to respect spatial integrity through the application of topology rules such as 'polygons must not overlap'. Vector data can also be used to represent continuously varying phenomena. Contour lines and triangulated irregular networks (TIN) are used to represent elevation or other continuously changing values.

TINs record values at point locations, which are connected by lines to form an irregular mesh of triangles. The face of the triangles represent the terrain surface.

Advantages and Disadvantages

There are some important advantages and disadvantages to using a raster or vector data model to represent reality:

- Raster datasets record a value for all points in the area covered which may require more storage space than representing data in a vector format that can store data only where needed.
- Raster data is computationally less expensive to render than vector graphics
- There are transparency and aliasing problems when overlaying multiple stacked pieces of raster images
- Vector data allows for visually smooth and easy implementation of overlay operations, especially in terms of graphics and shape-driven information like maps, routes and custom fonts, which are more difficult with raster data.
- Vector data can be displayed as vector graphics used on traditional maps, whereas raster data will appear as an image that may have a blocky appearance for object boundaries. (depending on the resolution of the raster file)
- Vector data can be easier to register, scale, and re-project, which can simplify combining vector layers from different sources.
- Vector data is more compatible with relational database environments, where they can be part of a relational table as a normal column and processed using a multitude of operators.
- Vector file sizes are usually smaller than raster data, which can be tens, hundreds or more times larger than vector data (depending on resolution).
- Vector data is simpler to update and maintain, whereas a raster image will have to be completely reproduced. (Example: a new road is added).
- Vector data allows much more analysis capability, especially for "networks" such as roads, power, rail, telecommunications, etc. (Examples: Best route, largest port, airfields connected to two-lane highways). Raster data will not have all the characteristics of the features it displays.

Non-spatial Data

Additional non-spatial data can also be stored along with the spatial data represented by the coordinates of a vector geometry or the position of a raster cell. In vector data, the additional data contains attributes of the feature. For example, a forest inventory polygon may also have an identifier value and information about tree species. In raster data the cell value can store attribute information, but it can also be used as an identifier that can relate to records in another table.

Software is currently being developed to support spatial and non-spatial decision-making, with the solutions to spatial problems being integrated with solutions to non-spatial problems. The end result with these flexible spatial decision-making support systems (FSDSSs) is expected to be that non-experts will be able to use GIS, along with spatial criteria, and simply integrate their non-spatial criteria to view solutions to multi-criteria problems. This system is intended to assist decision-making.

Popular GIS File Formats

Raster Formats

- ADRG – National Geospatial-Intelligence Agency (NGA)'s ARC Digitized Raster Graphics
- RPF – Raster Product Format, military file format specified in MIL-STD-2411
  - CADRG – Compressed ADRG, developed by NGA, nominal compression of 55:1 over ADRG (type of Raster Product Format)
  - CIB – Controlled Image Base, developed by NGA (type of Raster Product Format)
- Digital raster graphic (DRG) – digital scan of a paper USGS topographic map
- ECRG – National Geospatial-Intelligence Agency (NGA)'s Enhanced Compressed ARC Raster Graphics (Better resolution than CADRG and no colour loss)
- ECW – Enhanced Compressed Wavelet (from ERDAS). A compressed wavelet format, often lossy.
- Esri grid – proprietary binary and metadataless ASCII raster formats used by Esri
- GeoTIFF – TIFF variant enriched with GIS relevant metadata
- IMG – ERDAS IMAGINE image file format
- JPEG2000 – Open-source raster format. A compressed format, allows both lossy and lossless compression.

- MrSID – Multi-Resolution Seamless Image Database (by Lizardtech). A compressed wavelet format, allows both lossy and lossless compression.
- netCDF-CF – netCDF file format with CF medata conventions for earth science data. Binary storage in open format with optional compression. Allows for direct web-access of subsets/ aggregations of maps through OPeNDAP protocol.

Vector Formats

- AutoCAD DXF – contour elevation plots in AutoCAD DXF format (by Autodesk)
- Cartesian coordinate system (XYZ) – simple point cloud
- Digital Line Graph (DLG) – a USGS format for vector data
- Geography Markup Language (GML) – XML based open standard (by OpenGIS) for GIS data exchange
- GeoJSON – a lightweight format based on JSON, used by many open source GIS packages
- GeoMedia – Intergraph's Microsoft Access based format for spatial vector storage
- ISFC – Intergraph's MicroStation based CAD solution attaching vector elements to a relational Microsoft Access database
- Keyhole Markup Language (KML) – XML based open standard (by OpenGIS) for GIS data exchange
- MapInfo TAB format – MapInfo's vector data format using TAB, DAT, ID and MAP files
- National Transfer Format (NTF) – National Transfer Format (mostly used by the UK Ordnance Survey)
- Spatialite – is a spatial extension to SQLite, providing vector geodatabase functionality. It is similar to PostGIS, Oracle Spatial, and SQL Server with spatial extensions
- Shapefile – a popular vector data GIS format, developed by Esri
- Simple Features – Open Geospatial Consortium specification for vector data
- SOSI – a spatial data format used for all public exchange of spatial data in Norway
- Spatial Data File – Autodesk's high-performance geodatabase format, native to MapGuide
- TIGER – Topologically Integrated Geographic Encoding and Referencing

- Vector Product Format (VPF) – National Geospatial-Intelligence Agency (NGA)'s format of vectored data for large geographic databases

Grid Formats (for Elevation)

- USGS DEM – The USGS' Digital Elevation Model
  - GTOPO30 – Large complete Earth elevation model at 30 arc seconds, delivered in the USGS DEM format
- DTED – National Geospatial-Intelligence Agency (NGA)'s Digital Terrain Elevation Data, the military standard for elevation data
- GeoTIFF – TIFF variant enriched with GIS relevant metadata
- SDTS – The USGS' successor to DEM

Other Formats

- Dual Independent Map Encoding (DIME) – A historic GIS file format, developed in the 1960s
- Well-known text (WKT) – A text markup language for representing feature geometry, developed by Open Geospatial Consortium.
- Well-known binary (WKB) – Binary version of Well-known text.
- World file – Georeferencing a raster image file (e.g. JPEG, BMP)

Applications

The implementation of a GIS is often driven by jurisdictional (such as a city), purpose, or application requirements. Generally, a GIS implementation may be custom-designed for an organisation. Hence, a GIS deployment developed for an application, jurisdiction, enterprise, or purpose may not be necessarily interoperable or compatible with a GIS that has been developed for some other application, jurisdiction, enterprise, or purpose.

GIS provides, for every kind of location based organisation, a platform to update geographical data without wasting time to visit the field and update a database manually. GIS when integrated with other powerful enterprise solutions like SAP, helps creating powerful decision support system at enterprise level.

Many disciplines can benefit from GIS technology. An active GIS market has resulted in lower costs and continual improvements in the hardware and software components of GIS, and usage in the fields of science, government, business, and industry, with applications including real estate, public health, crime mapping, national defence, sustainable development, natural resources, climatology, landscape architecture, archaeology, regional and community planning, transportation and logistics.

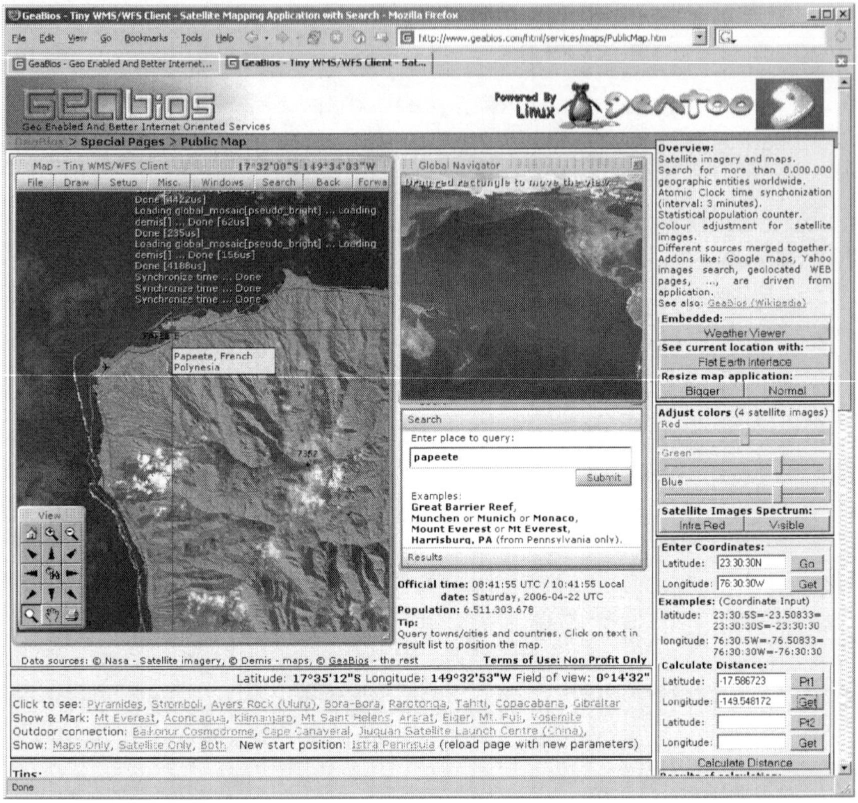

Figure: *GeaBios – tiny WMS/WFS client (Flash/DHTML)*

GIS is also diverging into location-based services, which allows GPS-enabled mobile devices to display their location in relation to fixed objects (nearest restaurant, gas station, fire hydrant) or mobile objects (friends, children, police car), or to relay their position back to a central server for display or other processing.

Open Geospatial Consortium Standards

The Open Geospatial Consortium (OGC) is an international industry consortium of 384 companies, government agencies, universities, and individuals participating in a consensus process to develop publicly available geoprocessing specifications. Open interfaces and protocols defined by OpenGIS Specifications support interoperable solutions that "geo-enable" the Web, wireless and location-based services, and mainstream IT, and empower technology developers to make complex spatial information and services accessible and useful with all kinds of applications. Open Geospatial Consortium protocols include Web Map Service, and Web Feature Service.

GIS products are broken down by the OGC into two categories, based on how completely and accurately the software follows the OGC specifications.

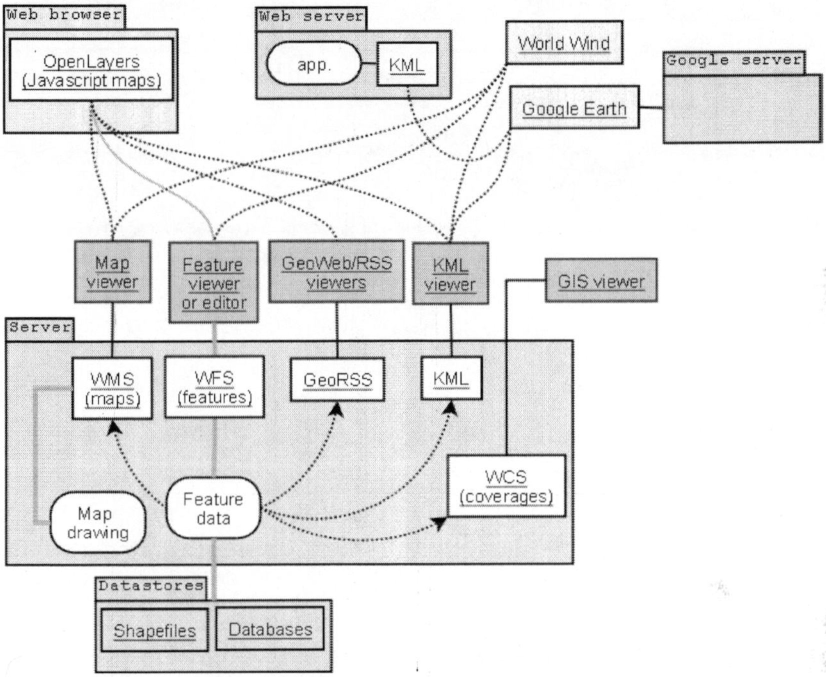

Figure: *OGC standards help GIS tools communicate.*

Compliant Products are software products that comply to OGC's OpenGIS Specifications. When a product has been tested and certified as compliant through the OGC Testing Program, the product is automatically registered as "compliant" on this site.

Implementing Products are software products that implement OpenGIS Specifications but have not yet passed a compliance test. Compliance tests are not available for all specifications. Developers can register their products as implementing draft or approved specifications, though OGC reserves the right to review and verify each entry.

Web Mapping

Web mapping is the process of using maps delivered by geographical information systems (GIS). Since a web map on the World Wide Web is both served and consumed, web mapping is more than just web cartography, it is both a service activity and consumer activity. Web GIS emphasizes geodata processing aspects more involved with design aspects such as data acquisition and server software architecture such as data storage and algorithms, than it does the end-user reports themselves. The terms *web GIS* and *web mapping* remain somewhat synonymous. Web GIS uses web maps, and end users who are *web mapping* are gaining analytical capabilities. The term *location-based services* refers to *web mapping* consumer goods and services. Web mapping usually involves a web browser or other user agent capable of client-server interactions.

While web mapping today is still being developed, challenges and innovations involving the feedback of the quality, the usability, the social benefits, and the legal constraints, drive its evolution.

Development and Implementation

The advent of web mapping can be regarded as a major new trend in cartography. Until recently cartography was restricted to a few companies, institutes and mapping agencies, requiring relatively expensive and complex hardware and software as well as skilled cartographers and geomatics engineers.

With the rise of web mapping, a range of data and technology was born - from free data generated by OpenStreetMap to proprietary datasets owned by Navteq, Google, Waze, and others. A range of free software to generate maps has also been conceived and implemented alongside proprietary tools like ArcGIS. As a result, the barrier to entry for serving maps on the web has been lowered.

Types of Web Maps

A first classification of web maps has been made by Kraak in 2001. He distinguished *static* and *dynamic* web maps and further distinguished *interactive* and *view only* web maps. Today there an increased number of dynamic web maps types, and static web map sources.

Analytic Web Maps

Analytic web maps offer GIS analysis. The geodata can be a static provision, or need updates. The borderline between analytic web maps and web GIS is fuzzy. Parts of the analysis can be carried out by the GIS geodata server. As web clients gain capabilities processing is distributed.

Animated Web Maps

Animated Maps show changes in the map over time by animating one of the graphical or temporal variables. Technologies enabling client-side display of animated web maps in include scalable vector graphics (SVG), Adobe Flash, Java, QuickTime, and others. Examples of animated web maps include weather maps and traffic congestion maps.

CartoDB launched an open source library, Torque, which enables the creation of dynamic animated maps with millions of records. Twitter uses this technology to create maps to reflect how users reacted to news and events worldwide.

In recent years there has been an explosion of mapping applications on the web such as Google Maps and Bing Maps. These websites give the public access to huge amounts of geographic data.

Some of them, like Google Maps and OpenLayers, expose an API that enable users to create custom applications. These toolkits commonly offer street maps, aerial/satellite imagery, geocoding, searches, and routing functionality. Web mapping has also uncovered the potential of crowd sourcing geodata in projects like OpenStreetMap, which is a collaborative project to create a free editable map of the world.

Adding the Dimension of Time

The condition of the Earth's surface, atmosphere, and subsurface can be examined by feeding satellite data into a GIS. GIS technology gives researchers the ability to examine the variations in Earth processes over days, months, and years. As an example, the changes in vegetation vigor through a growing season can be animated to determine when drought was most extensive in a particular region. The resulting graphic, known as a normalized vegetation index, represents a rough measure of plant health. Working with two variables over time would then allow researchers to detect regional differences in the lag between a decline in rainfall and its effect on vegetation.

GIS technology and the availability of digital data on regional and global scales enable such analyses. The satellite sensor output used to generate a vegetation graphic is produced for example by the Advanced Very High Resolution Radiometre (AVHRR). This sensor system detects the amounts of energy reflected from the Earth's surface across various bands of the spectrum for surface areas of about 1 square kilometre. The satellite sensor produces images of a particular location on the Earth twice a day. AVHRR and more recently the Moderate-Resolution Imaging Spectroradiometre (MODIS) are only two of many sensor systems used for Earth surface analysis. More sensors will follow, generating ever greater amounts of data.

In addition to the integration of time in environmental studies, GIS is also being explored for its ability to track and model the progress of humans throughout their daily routines. A concrete example of progress in this area is the recent release of time-specific population data by the U.S. Census. In this data set, the populations of cities are shown for daytime and evening hours highlighting the pattern of concentration and dispersion generated by North American commuting patterns. The manipulation and generation of data required to produce this data would not have been possible without GIS.

Using models to project the data held by a GIS forward in time have enabled planners to test policy decisions using spatial decision support systems.

Semantics

Tools and technologies emerging from the W3C's Data Activity are proving useful for data integration problems in information systems. Correspondingly, such technologies have been proposed as a means to facilitate interoperability and data reuse among GIS applications and also to enable new analysis mechanisms.

Ontologies are a key component of this semantic approach as they allow a formal, machine-readable specification of the concepts and relationships in a given domain. This in turn allows a GIS to focus on the intended meaning of data rather than its syntax or structure. For example, reasoning that a land cover type classified as *deciduous needleleaf trees* in one dataset is a specialisation or subset of land cover type *forest* in another more roughly classified dataset can help a GIS automatically merge the two datasets under the more general land cover classification. Tentative ontologies have been developed in areas related to GIS applications, for example the hydrology ontology developed by the Ordnance Survey in the United Kingdom and the SWEET ontologies developed by NASA's Jet Propulsion Laboratory. Also, simpler ontologies and semantic metadata standards are being proposed by the W3C Geo Incubator Group to represent geospatial data on the web. GeoSPARQL is a standard developed by the Ordnance Survey, United States Geological Survey, Natural Resources Canada, Australia's Commonwealth Scientific and Industrial Research Organisation and others to support ontology creation and reasoning using well-understood OGC literals (GML, WKT), topological relationships (Simple Features, RCC8, DE-9IM), RDF and the SPARQL database query protocols.

Exploring Spatial Data Visually

Map

A map is a symbolic depiction highlighting relationships between elements of some space, such as objects, regions, and themes. Many maps are static two-dimensional, geometrically accurate (or approximately accurate) representations of three-dimensional space, while others are dynamic or interactive, even three-dimensional. Although most commonly used to depict geography, maps may represent any space, real or imagined, without regard to context or scale; e.g. brain mapping, DNA mapping and extraterrestrial mapping.

Although the earliest maps known are of the heavens, geographic maps of territory have a very long tradition and exist from ancient times. The word "map" comes from the medieval Latin *Mappa mundi*, wherein *mappa* meant napkin or cloth and *mundi* the world. Thus, "map" became the shortened term referring to a two-dimensional representation of the surface of the world.

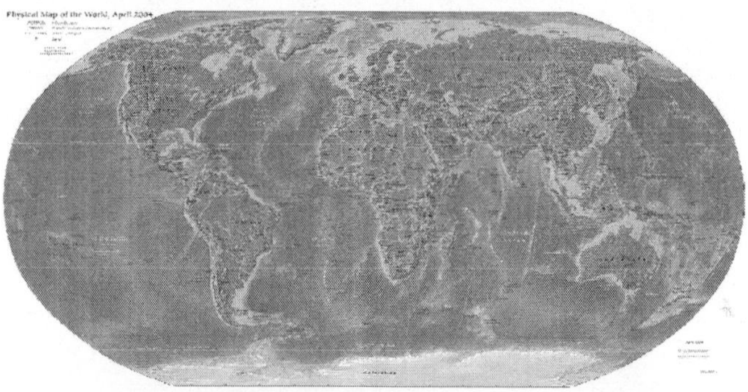

Figure: *World map*

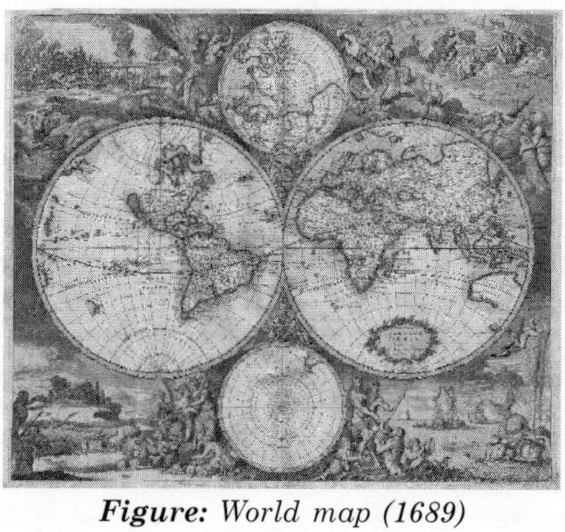

Figure: *World map (1689)*

Geographic Maps

Figure: *A celestial map from the 17th century, by the cartographer Frederik de Wit.*

Cartography or *map-making* is the study and practice of crafting representations of the Earth upon a flat surface, and one who makes maps is called a cartographer.

Road maps are perhaps the most widely used maps today, and form a subset of navigational maps, which also include aeronautical and nautical charts, railroad network maps, and hiking and bicycling maps. In terms of quantity, the largest number of drawn map sheets is probably made up by local surveys, carried out by municipalities, utilities, tax assessors, emergency services providers, and other local agencies. Many national surveying projects have been carried out by the military, such as the British Ordnance Survey: a civilian government agency, internationally renowned for its comprehensively detailed work.

In addition to location information maps may also be used to portray contour lines indicating constant values of elevation, temperature, rainfall, etc.

Orientation of Maps

The orientation of a map is the relationship between the directions on the map and the corresponding compass directions in reality. The word "orient" is derived from Latin *oriens*, meaning East. In the Middle Ages many maps, including the T and O maps, were drawn with East at the top (meaning that the direction "up" on the map corresponds to East on the compass). Today, the most common – but far from universal – cartographic convention is that North is at the top of a map.

Figure: The Hereford Mappa Mundi, *about 1300, Hereford Cathedral, England. A classic "T-O" map with Jerusalem at centre, east toward the top, Europe the bottom left and Africa on the right.*

Several kinds of maps are often traditionally not oriented with North at the top:

- Maps from non-Western traditions are oriented a variety of ways. Old maps of Edo show the Japanese imperial palace as the "top", but also at the centre, of the map. Labels on the map are oriented in such a way that you cannot read them properly unless you put the imperial palace above your head.
- Medieval European T and O maps such as the Hereford Mappa Mundi were centred on Jerusalem with East at the top. Indeed, prior to the reintroduction of Ptolemy's *Geography* to Europe around 1400, there was no single convention in the West. Portolan charts, for example, are oriented to the shores they describe.
- Maps of cities bordering a sea are often conventionally oriented with the sea at the top.
- Route and channel maps have traditionally been oriented to the road or waterway they describe.

- Polar maps of the Arctic or Antarctic regions are conventionally centred on the pole; the direction North would be towards or away from the centre of the map, respectively. Typical maps of the Arctic have 0° meridian towards the bottom of the page; maps of the Antarctic have the 0° meridian towards the top of the page.
- Reversed maps, also known as Upside-Down maps or South-Up maps, reverse the "North is up" convention and have South at the top.
- Buckminster Fuller's Dymaxion maps are based on a projection of the Earth's sphere onto an icosahedron. The resulting triangular pieces may be arranged in any order or orientation.
- Modern digital GIS maps such as ArcMap typically project north at the top of the map, but use math degrees (0 is east, degrees increase counter-clockwise), rather than compass degrees (0 is north, degrees increase clockwise) for orientation of transects. Compass decimal degrees can be converted to math degrees by subtracting them from 450; if the answer is greater than 360, subtract 360.

Scale and Accuracy

Many, but not all, maps are drawn to a scale, expressed as a ratio such as 1:10,000, meaning that 1 of any unit of measurement on the map corresponds exactly, to 10,000 of that same unit on the ground. The scale statement may be taken as exact when the region mapped is small enough for the curvature of the Earth to be neglected, for example in a town planner's city map. Over larger regions where the curvature cannot be ignored we must use map projections from the curved surface of the Earth (sphere or ellipsoid) to the plane. The impossibility of flattening the sphere to the plane implies that no map projection can have constant scale: on most projections the best we can achieve is accurate scale on one or two lines (not necessarily straight) on the projection. Thus for map projections we must introduce the concept of point scale, which is a function of position, and strive to keep its variation within narrow bounds. Although the scale statement is nominal it is usually accurate enough for all but the most precise of measurements.

Large scale maps, say 1:10,000, cover relatively small regions in great detail and small scale maps, say 1:10,000,000, cover large regions such as nations, continents and the whole globe. The large/small terminology arose from the practice of writing scales as numerical fractions: 1/10,000 is larger than 1/10,000,000. There is no exact dividing line between large and small but 1/100,000 might well be considered as a medium scale. Examples of large scale maps are the 1:25,000 maps produced for hikers; on the other hand maps intended for motorists at 1:250,000 or 1:1,000,000 are small scale.

It is important to recognise that even the most accurate maps sacrifice a certain amount of accuracy in scale to deliver a greater visual usefulness to its user. For example, the width of roads and small streams are exaggerated when they are too narrow to be shown on the map at true scale; that is, on a printed map they would be narrower than could be perceived by the naked eye. The same applies to computer maps where the smallest unit is the pixel. A narrow stream say must be shown to have the width of a pixel even if at the map scale it would be a small fraction of the pixel width.

Some maps, called cartograms, have the scale deliberately distorted to reflect information other than land area or distance. For example, this map (at the right) of Europe has been distorted to show population distribution, while the rough shape of the continent is still discernible.

Another example of distorted scale is the famous London Underground map. The basic geographical structure is respected but the tube lines (and the River Thames) are smoothed to clarify the relationships between stations. Near the centre of the map stations are spaced out more than near the edges of map.

Further inaccuracies may be deliberate. For example, cartographers may simply omit military installations or remove features solely in order to enhance the clarity of the map. For example, a road map may not show railroads, smaller waterways or other prominent non-road objects, and even if it does, it may show them less clearly (e.g. dashed or dotted lines/outlines) than the main roads. Known as decluttering, the practice makes the subject matter that the user is interested in easier to read, usually without sacrificing overall accuracy. Software-based maps often allow the user to toggle decluttering between ON, OFF and AUTO as needed. In AUTO the degree of decluttering is adjusted as the user changes the scale being displayed.

Map Types and Projections

Maps of the world or large areas are often either 'political' or 'physical'. The most important purpose of the political map is to show territorial borders; the purpose of the physical is to show features of geography such as mountains, soil type or land use including infrastructure such as roads, railroads and buildings. Topographic maps show elevations and relief with contour lines or shading. Geological maps show not only the physical surface, but characteristics of the underlying rock, fault lines, and subsurface structures. Maps that depict the surface of the Earth also use a projection, a way of translating the three-dimensional real surface of the geoid to a two-dimensional picture. Perhaps the best-known world-map projection is the Mercator projection, originally designed as a form of nautical chart. Aeroplane pilots use aeronautical charts based on a Lambert conformal conic projection, in which a cone is laid over the section of the earth to be mapped. The cone intersects the sphere (the earth) at one or two parallels which are chosen as standard lines. This allows the pilots to plot a great-circle route approximation on a flat, two-dimensional chart.

- Azimuthal or Gnomonic map projections are often used in planning air routes due to their ability to represent great circles as straight lines.
- Richard Edes Harrison produced a striking series of maps during and after World War II for Fortune magazine. These used "bird's eye" projections to emphasise globally strategic "fronts" in the air age, pointing out proximities and barriers not apparent on a conventional rectangular projection of the world.

Electronic Maps

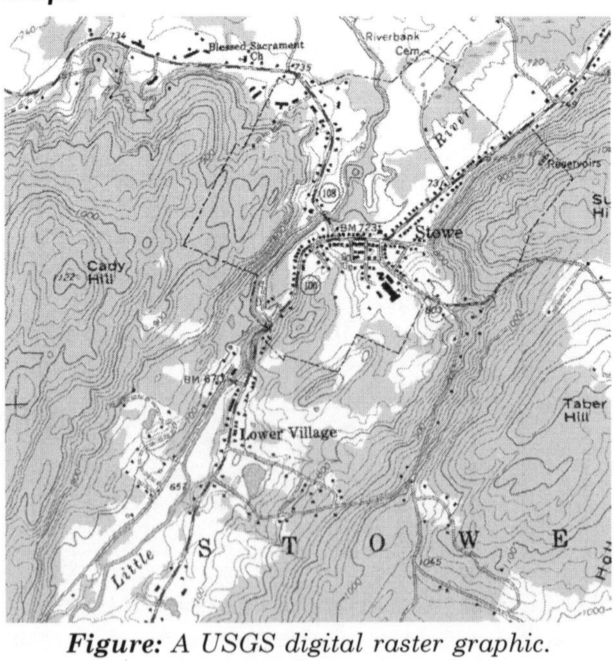

Figure: *A USGS digital raster graphic.*

From the last quarter of the 20th century, the indispensable tool of the cartographer has been the computer. Much of cartography, especially at the data-gathering survey level, has been subsumed by Geographic Information Systems (GIS). The functionality of maps has been greatly advanced by technology simplifying the superimposition of spatially located variables onto existing geographical maps. Having local information such as rainfall level, distribution of wildlife, or demographic data integrated within the map allows more efficient analysis and better decision making. In the pre-electronic age such superimposition of data led Dr. John Snow to identify the location of an outbreak of cholera. Today, it is used by agencies of the human kind, as diverse as wildlife conservationists and militaries around the world.

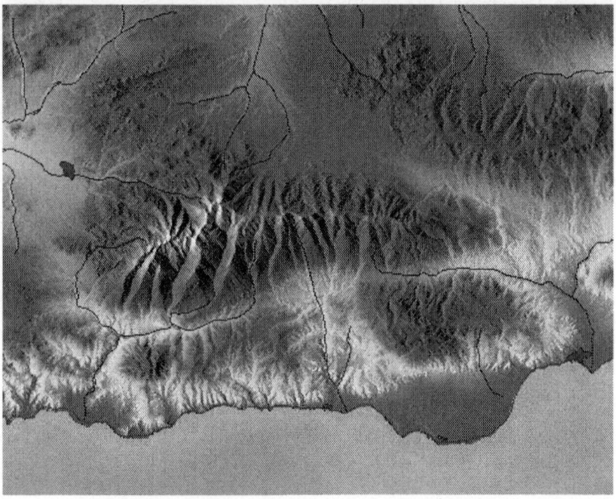

Figure: *Relief map Sierra Nevada*

Even when GIS is not involved, most cartographers now use a variety of computer graphics programs to generate new maps.

Interactive, computerised maps are commercially available, allowing users to *zoom in* or *zoom out* (respectively meaning to increase or decrease the scale), sometimes by replacing one map with another of different scale, centred where possible on the same point. In-car global navigation satellite systems are computerised maps with route-planning and advice facilities which monitor the user's position with the help of satellites. From the computer scientist's point of view, zooming in entails one or a combination of:

1. replacing the map by a more detailed one
2. enlarging the same map without enlarging the pixels, hence showing more detail by removing less information compared to the less detailed version

3. enlarging the same map with the pixels enlarged (replaced by rectangles of pixels); no additional detail is shown, but, depending on the quality of one's vision, possibly more detail can be seen; if a computer display does not show adjacent pixels really separate, but overlapping instead (this does not apply for an LCD, but may apply for a cathode ray tube), then replacing a pixel by a rectangle of pixels does show more detail. A variation of this method is interpolation.

For example:

- Typically (2) applies to a Portable Document Format (PDF) file or other format based on vector graphics. The increase in detail is, of course, limited to the information contained in the file: enlargement of a curve may eventually result in a series of standard geometric figures such as straight lines, arcs of circles or splines.
- (2) may apply to text and (3) to the outline of a map feature such as a forest or building.
- (1) may apply to the text as needed (displaying labels for more features), while (2) applies to the rest of the image. Text is not necessarily enlarged when zooming in. Similarly, a road represented by a double line may or may not become wider when one zooms in.
- The map may also have layers which are partly raster graphics and partly vector graphics. For a single raster graphics image (2) applies until the pixels in the image file correspond to the pixels of the display, thereafter (3) applies.

Conventional Signs

The various features shown on a map are represented by conventional signs or symbols. For example, colours can be used to indicate a classification of roads. Those signs are usually explained in the margin of the map, or on a separately published characteristic sheet.

Some cartographers prefer to make the map cover practically the entire screen or sheet of paper, leaving no room "outside" the map for information about the map as a whole. These cartographers typically place such information in an otherwise "blank" region "inside" the map — cartouche, map legend, title, compass rose, bar scale, etc. In particular, some maps contain smaller "sub-maps" in otherwise blank regions—often one at a much smaller scale showing the whole globe and where the whole map fits on that globe, and a few showing "regions of interest" at a larger scale in order to show details that wouldn't otherwise fit.

Occasionally sub-maps use the same scale as the large map—a few maps of the contiguous United States include a sub-map to the same scale for each of the two non-contiguous states.

Labelling

To communicate spatial information effectively, features such as rivers, lakes, and cities need to be labelled. Over centuries cartographers have developed the art of placing names on even the densest of maps. Text placement or name placement can get mathematically very complex as the number of labels and map density increases. Therefore, text placement is time-consuming and labour-intensive, so cartographers and GIS users have developed automatic label placement to ease this process.

Non-Geographical Spatial Maps

Maps exist of the solar system, and other cosmological features such as star maps. In addition maps of other bodies such as the Moon and other planets are technically not *geo*graphical maps.

Non Spatial Maps

Diagrams such as schematic diagrams and Gantt charts and treemaps display logical relationships between items, and do not display spatial relationships at all.

Some maps, for example the London Underground map, are topological maps. Topological in nature, the distances are completely unimportant; only the connectivity is significant.

General-purpose Maps

General-purpose maps provide many types of information on one map. Most atlas maps, wall maps, and road maps fall into this category. The following are some features that might be shown on a general-purpose maps: bodies of water, roads, railway lines, parks, elevations, towns and cities, political boundaries, latitude and longitude, national and provincial parks. These maps give a broad understanding of location and features of an area. You can gain an understanding of the type of landscape, the location of urban places, and the location of major transportation routes all at once.

Digital Geologic Mapping

Screenshot of a structure map generated by geological mapping software for an 8500ft deep gas & Oil reservoir in the Erath field, Vermilion Parish, Erath, Louisiana. The left-to-right gap, near the top of the contour map indicates a Fault line. This fault line is between the blue/green contour lines and the purple/red/yellow contour lines. The thin red circular contour line in the middle of the map indicates the top of the oil reservoir. Because gas floats above oil, the thin red contour line marks the gas/oil contact zone.

Digital geologic mapping is the process by which geologic features are observed, analyzed, and recorded in the field and displayed in real-time on a computer or personal digital assistant (PDA). The primary function of this emerging technology is to produce spatially referenced geologic maps that can be utilized and updated while conducting field work.

Traditional Geologic Mapping

Geologic mapping is an interpretive process involving multiple types of information, from analytical data to personal observation, all synthesized and recorded by the geologist. Geologic observations have traditionally been recorded on paper, whether on standardized note cards, in a notebook, or on a map.

Mapping in the Digital Era

In the 21st century, computer technology and software are becoming portable and powerful enough to take on some of the more mundane tasks a geologist must perform in the field, such as precisely locating oneself with a GPS unit, displaying multiple images (maps, satellite images, aerial photography, etc.), plotting strike and dip symbols, and colour-coding different physical characteristics of a lithology or contact type (e.g., unconformity) between rock strata. Additionally, computers can now perform some tasks that were difficult to accomplish in the field, for example, handwriting or voice recognition and annotating photographs on the spot.

Digital mapping has positive and negative effects on the mapping process; only an assessment of its impact on a geological mapping project as a whole shows whether it provides a net benefit. With the use of computers in the field, the recording of observations and basic data management changes dramatically. The use of digital mapping also affects when data analysis occurs in the mapping process, but does not greatly affect the process itself.

Some Advantages of Digital Mapping

- Data entered by a geologist may have fewer errors than data transcribed by a data entry clerk.
- Data entry by geologists in the field may take less total time than subsequent data entry in the office, potentially reducing the overall time needed to complete a project.

- The spatial extent of real world objects and their attributes can be entered directly into a database with geographic information system (GIS) capability. Features can be automatically colour-coded and symbolized based on set criteria.
- Multiple maps and imagery (geophysical maps, satellite images, orthophotos, etc.) can easily be carried and displayed on-screen.
- Geologists may upload each other's data files for the next day's field work as reference.
- Data analysis may start immediately after returning from the field, since the database has already been populated.
- Data can be constrained by dictionaries and dropdown menus to ensure that data are recorded systematically and that mandatory data are not forgotten
- Labour-saving tools and functionality can be provided in the field e.g. structure contours on the fly, and 3D visualisation
- Systems can be wirelessly connected to other digital field equipment (such as digital cameras and sensor webs)

Some Disadvantages of Digital Mapping

- Computers and related items (extra batteries, stylus, cameras, etc.) must be carried in the field.
- Field data entry into the computer may take longer than physically writing on paper, possibly resulting in longer field programs.
- Data entered by multiple geologists may contain more inconsistencies than data entered by one person, making the database more difficult to query.
- Written descriptions convey to the reader detailed information through imagery that may not be communicated by the same data in parsed format.
- Geologists may be inclined to shorten text descriptions because they are difficult to enter (either by handwriting or voice recognition), resulting in loss of data.
- There are no original, hardcopy field maps or notes to archive. Paper is a more stable medium than digital format.

Educational and scientific uses

Some universities and secondary educators are integrating digital geologic mapping into class work. For example, The GeoPad project describes the combination of technology, teaching field geology, and geologic mapping in programs such as Bowling Green State University's geology field camp. At Urbino University (Italy) it:Università di Urbino, Field Digital Mapping Techniques are integrated in Earth and Environmental Sciences courses since 2006. The MapTeach program is designed to provide hands-on digital mapping for middle and high school students. The SPLINT project in the UK is using the BGS field mapping system as part of their teaching curriculum

Digital mapping technology can be applied to traditional geologic mapping, reconnaissance mapping, and surveying of geologic features. At international digital field data capture (DFDC) meetings, major geological surveys (e.g., British Geological Survey and Geological Survey of Canada) discuss how to harness and develop the technology. Many other geological surveys and private companies are also designing systems to conduct scientific and applied geological mapping of, for example, geothermal springs and mine sites.

Digital Mapping Equipment

The initial cost of digital geologic computing and supporting equipment may be significant. In addition, equipment and software must be replaced occasionally due to damage, loss, and obsolescence. Products moving through the market are quickly discontinued as technology and consumer interests evolve. A product that works well for digital mapping may not be available for purchase the following year; however, testing multiple brands and generations of equipment and software is prohibitively expensive.

Common Essential Features

Some features of digital mapping equipment are common to both survey or reconnaissance mapping and "traditional" comprehensive mapping. The capture of less data-intensive reconnaissance mapping or survey data in the field can be accomplished by less robust databases and GIS programs, and hardware with a smaller screen size.

- Devices and software are intuitive to learn and easy to use
- Rugged, as typically defined by military standards (MIL-STD-810) and ingress protection ratings
- Waterproof
- Screen is easy to read in bright sunlight and on gray sky days
- Removable static memory cards can be used to back up data
- Memory on board is recoverable
- Real-time and post-processing differential correction for GPS locations

- Portable battery with at least 9 hours of life at near constant use
- Can change batteries in the field
- Batteries should have no "memory," such as with NiCd
- Chargeable by unconventional power sources (generators, solar, etc.)
- Wireless real-time link to GPS or built-in GPS
- Wireless real-time link from computer to camera and other peripherals
- USB port(s)

Features Essential to Capture Traditional Geologic Observations

Hardware and software only recently (in 2000) became available that can satisfy most of the criteria necessary for digitally capturing "traditional" mapping data.

- Screen about 5" x 7"—compact but large enough to see map features. In 2009, some traditional mapping is conducted on PDAs.
- Lightweight—ideally less than 3 lbs.
- Transcription to digital text from handwriting and voice recognition.
- Can store paragraphs of data (text fields).
- Can store complex relational database with drop-down lists.
- Operating system and hardware are compatible with a robust GIS program.
- At least 512 MB memory.

Technology

Software

Since every geologic mapping project covers an area with unique lithologies and complexities, and every geologist has a unique style of mapping, no software is perfect for digital geologic mapping out of the box. The geologist can chose to either modify their mapping style to the available software, or modify the software to their mapping style, which may require extensive programming. As of 2009, available geologic mapping software (ESRI's ArcPad, PenMap, BeeGIS require some degree of customization for a given geologic mapping project. Some digital-mapping geologists/programmers have chosen to highly customize or extend ESRI's ArcGIS instead (e.g., BGS-SIGMAmobile and GeoRover. At digital field data capture meetings such as at the British Geological Survey in 2002 some organisations agreed to share development experiences, and some software systems are now available to download for free, such as BGS-SIGMAmobile.

Hardware

A computer or PDA, GPS, stylus, and sometimes a rangefinder are required for digital geologic mapping. As of 2009, ruggedized tablet PCs and laptops (e.g., Trimble Navigation's Yuma, Xplore Technologies' iX104 series, Group Mobile's Getac series are being used in the field. Geologists needing a lighter weight option with longer battery life use PDAs (e.g., Trimble Navigation's GeoExplorer series instead. GPS units are commonly built into the computers, but they may also be external and wireless.

Maps

- are limited to two-dimensions
  - o must show 3-D data projected onto a flat surface
  - o give a distorted impression of spatial distributions on the globe
- are static, cannot show change through time or animate
- have difficulty showing interactions or flows between places
- are limited by the tools used to make maps
  - o pens of constant width
  - o constant colour or tone
  - o the airbrush adds flexibility but is difficult to use, control
- have difficulty showing uncertainty in data
  - o give a false impression of accuracy

Computer-generated displays

- include screens, plots, printer output
- include raster and vector
- can be animated
- can show continuous gradations of colour, texture, tone
- can show 3-D using stereoscopic technology and pairs of images
- the computer is a powerful tool for visualizing spatial information
  - o this unit looks at some of the issues involved in combining the knowledge of cartography with the power of digital technology

- o all too often these issues are ignored when output maps and displays are created from GIS
- o although GIS display and mapping has much to learn from principles of cartographic design, it also provides entirely new possibilities

Cartographic Background

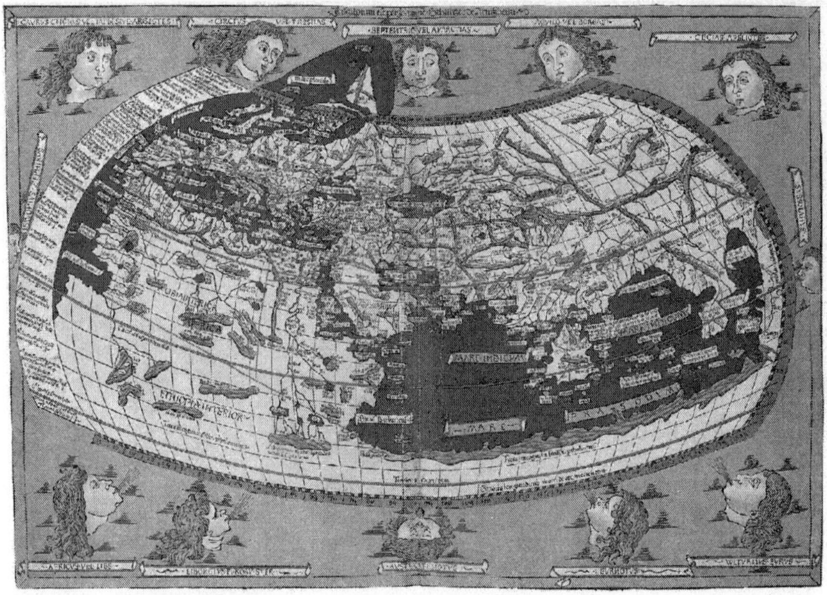

Figure: *A medieval depiction of the Ecumene (1482, Johannes Schnitzer, engraver), constructed after the coordinates in Ptolemy's Geography and using his second map projection. The translation into Latin and dissemination of Geography in Europe, in the beginning of the 15th century, marked the rebirth of scientific cartography, after more than a millennium of stagnation*

Cartography is the study and practice of making maps. Combining science, aesthetics, and technique, cartography builds on the premise that reality can be modelled in ways that communicate spatial information effectively.

The fundamental problems of traditional cartography are to:

- Set the map's agenda and select traits of the object to be mapped. This is the concern of map editing. Traits may be physical, such as roads or land masses, or may be abstract, such as toponyms or political boundaries.
- Represent the terrain of the mapped object on flat media. This is the concern of map projections.

- Eliminate characteristics of the mapped object that are not relevant to the map's purpose. This is the concern of generalization.
- Reduce the complexity of the characteristics that will be mapped. This is also the concern of generalization.
- Orchestrate the elements of the map to best convey its message to its audience. This is the concern of map design.

Modern cartography is largely integrated with geographic information science (GIScience) and constitutes many theoretical and practical foundations of geographic information systems.

History

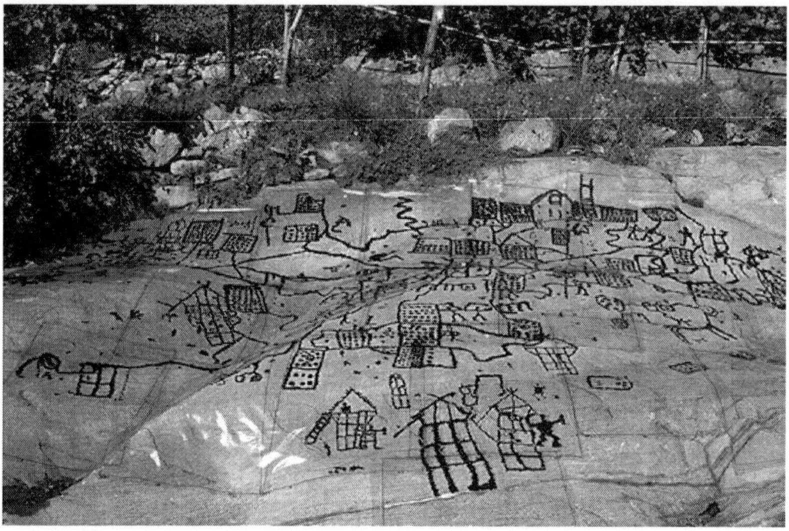

Figure: *The Bedolina Map and its tracing, VI-IV cent. BC*

The earliest known map is a matter of some debate, both because the definition of "map" is not sharp and because some artifacts speculated to be maps might actually be something else. A wall painting, which may depict the ancient Anatolian city of Çatalhöyük (previously known as Catal Huyuk or Çatal Hüyük), has been dated to the late 7th millennium BCE. Among the prehistoric alpine rock carvings of Mount Bego (F) and Valçamonica (I), geometric patterns (dotted rectangles and lines) are widely interpreted in archaeological literature as a plan depiction of cultivated plots. Defined as "topographic representations" and well dated to the IV millennium BC, they witness the introduction of the agriculture in the alpine territory. Other known maps of the ancient world include the Minoan "House of the Admiral" wall painting from c. 1600 BCE, showing a seaside community in an oblique perspective and an engraved map of the holy Babylonian city of Nippur, from the Kassite period (14th – 12th centuries BCE). The oldest surviving world maps are the Babylonian world maps from the 9th century BCE. One shows Babylon on the Euphrates, surrounded by a circular landmass showing Assyria, Urartu and several cities, in turn surrounded by a "bitter river" (Oceanus), with seven islands arranged around it. Another depicts Babylon as being further north from the centre of the world.

Topographic compositions were also carved in the alpine rock art during the Iron Age. Best known is the Bedolina Map, in Valcamonica, with a composition dated to the VI-IV cent. BC. The ancient Greeks and Romans created maps, beginning at latest with Anaximander in the 6th century BC. In the 2nd century AD, Ptolemy produced his treatise on cartography, Geographia. This contained Ptolemy's world map – the world then known to Western society *(Ecumene)*. As early as the 8th century, Arab scholars were translating the works of the Greek geographers into Arabic.

In ancient China, geographical literature spans back to the 5th century BC. The oldest extant Chinese maps come from the State of Qin, dated back to the 4th century BC, during the Warring States period. In the book of the *Xin Yi Xiang Fa Yao*, published in 1092 by the Chinese scientist Su Song, a star map on the equidistant cylindrical projection. Although this method of charting seems to have existed in China even prior to this publication and scientist, the greatest significance of the star maps by Su Song is that they represent the oldest existent star maps in printed form.

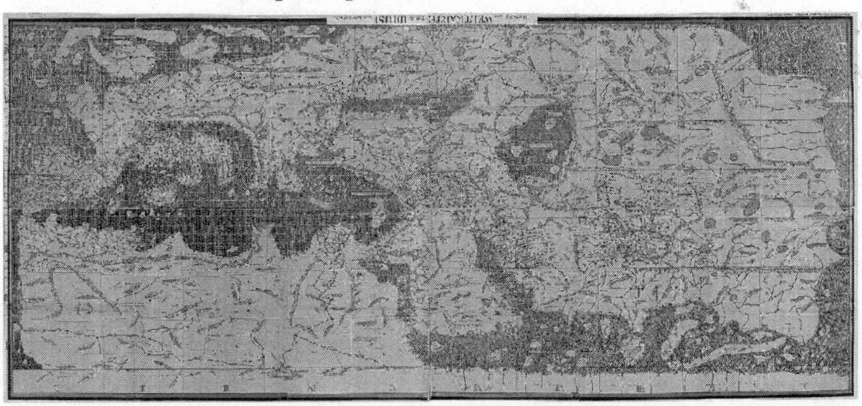

Figure: *The* Tabula Rogeriana, *drawn by Muhammad al-Idrisi for Roger II of Sicily in 1154*

Early forms of cartography of India included the locations of the Pole star and other constellations of use. These charts may have been in use by the beginning of the Common Era for purposes of navigation.

Mappa mundi are the Medieval European maps of the world. Approximately 1,100 mappae mundi are known to have survived from the Middle Ages. Of these, some 900 are found illustrating manuscripts and the remainder exist as stand-alone documents.

The Arab geographer Muhammad al-Idrisi produced his medieval atlas *Tabula Rogeriana* in 1154. He incorporated the knowledge of Africa, the Indian Ocean and the Far East, gathered by Arab merchants and explorers with the information inherited from the classical geographers to create the most accurate map of the world up until his time. It remained the most accurate world map for the next three centuries.

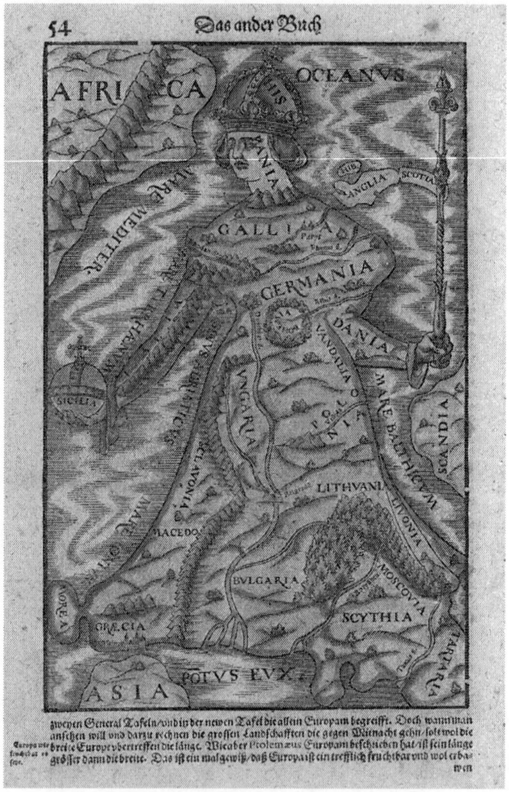

Figure: Europa regina *in Sebastian Münster's* "Cosmographia"*, 1570*

In the Age of Exploration, from the 15th century to the 17th century, European cartographers both copied earlier maps (some of which had been passed down for centuries) and drew their own based on explorers' observations and new surveying techniques. The invention of the magnetic compass, telescope and sextant enabled increasing accuracy. In 1492, Martin Behaim, a German cartographer, made the oldest extant globe of the Earth.

Johannes Werner refined and promoted the Werner projection. In 1507, Martin Waldseemüller produced a globular world map and a large 12-panel world wall map (*Universalis Cosmographia*) bearing the first use of the name "America". Portuguese cartographer Diego Ribero was the author of the first known planisphere with a graduated Equator (1527). Italian cartographer Battista Agnese produced at least 71 manuscript atlases of sea charts.

Due to the sheer physical difficulties inherent in cartography, map-makers frequently lifted material from earlier works without giving credit to the original cartographer. For example, one of the most famous early maps of North America is unofficially known as the "Beaver Map", published in 1715 by Herman Moll. This map is an exact reproduction of a 1698 work by Nicolas de Fer. De Fer in turn had copied images that were first printed in books by Louis Hennepin, published in 1697, and François Du Creux, in 1664. By the 18th century, map-makers started to give credit to the original engraver by printing the phrase "After [the original cartographer]" on the work.

Technological Changes

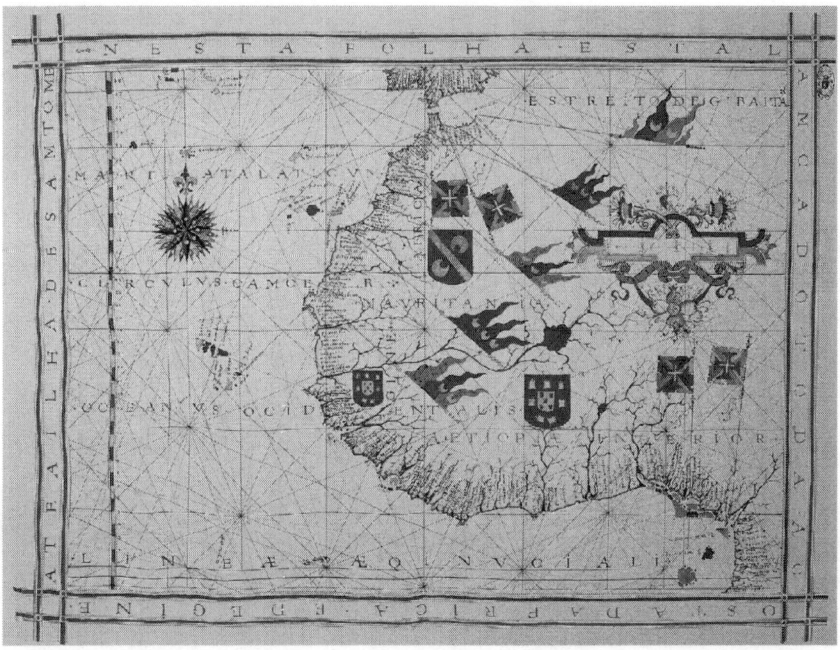

Figure: *A pre-Mercator nautical chart of 1571, from Portuguese cartographer Fernão Vaz Dourado (c. 1520–c. 1580). It belongs to the so-called* plane chart *model, where observed latitudes and magnetic directions are plotted directly into the plane, with a constant scale, as if the Earth were a plane (Portuguese National Archives of Torre do Tombo, Lisbon).*

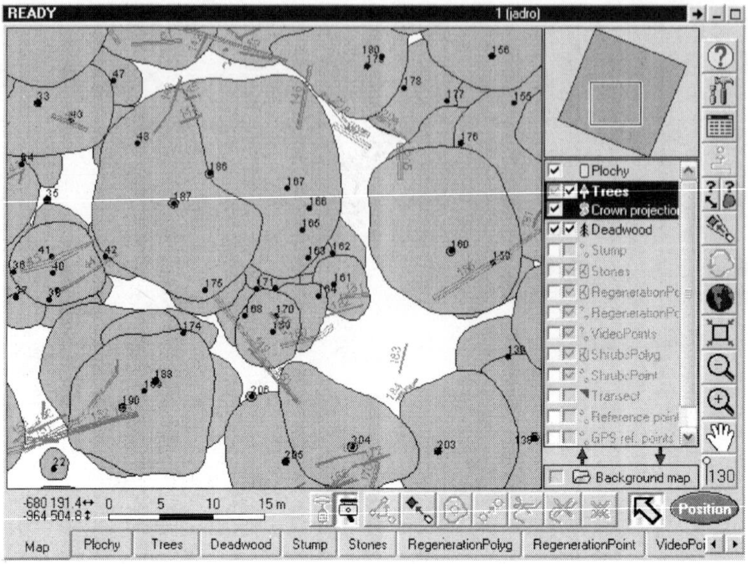

Figure: *Mapping can be done with GPS and laser rangefinder directly in the field. Image shows mapping of forest structure (position of trees, dead wood and canopy).*

In cartography, technology has continually changed in order to meet the demands of new generations of mapmakers and map users. The first maps were manually constructed with brushes and parchment; therefore, varied in quality and were limited in distribution. The advent of magnetic devices, such as the compass and much later, magnetic storage devices, allowed for the creation of far more accurate maps and the ability to store and manipulate them digitally.

Advances in mechanical devices such as the printing press, quadrant and vernier, allowed for the mass production of maps and the ability to make accurate reproductions from more accurate data. Optical technology, such as the telescope, sextant and other devices that use telescopes, allowed for accurate surveying of land and the ability of mapmakers and navigators to find their latitude by measuring angles to the North Star at night or the sun at noon.

Advances in photochemical technology, such as the lithographic and photochemical processes, have allowed for the creation of maps that have fine details, do not distort in shape and resist moisture and wear. This also eliminated the need for engraving, which further shortened the time it takes to make and reproduce maps.

In the 20th century, Aerial photography, satellite imagery, and remote sensing provided efficient, precise methods for mapping physical features, such as coastlines, roads, buildings, watersheds, and topography. Advancements in electronic technology ushered in another revolution in cartography. Ready availability of computers and peripherals such as monitors, plotters, printers, scanners (remote and document) and analytic stereo plotters, along with computer programs for visualization, image processing, spatial analysis, and database management, democratized and greatly expanded the making of maps. The ability to superimpose spatially located variables onto existing maps created new uses for maps and new industries to explore and exploit these potentials.

These days most commercial-quality maps are made using software that falls into one of three main types: CAD, GIS and specialised illustration software. Spatial information can be stored in a database, from which it can be extracted on demand. These tools lead to increasingly dynamic, interactive maps that can be manipulated digitally.

With the field rugged computers, GPS and laser rangefinders, it is possible to perform mapping directly in the terrain.

Map Types

General vs. Thematic Cartography

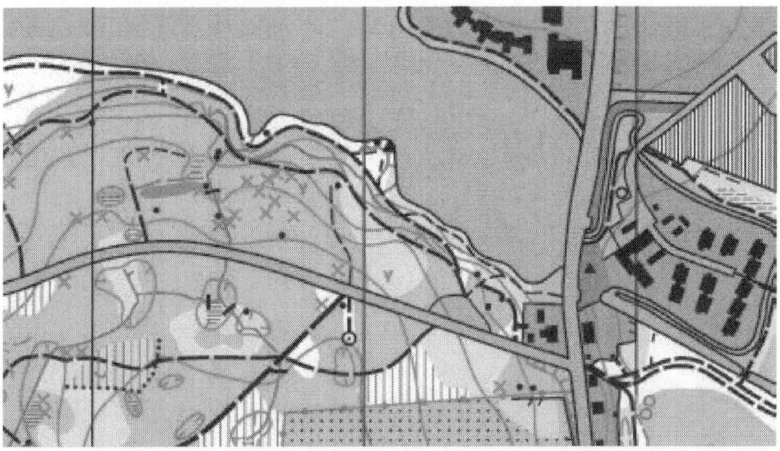

Figure: *Small section of an orienteering map.*

In understanding basic maps, the field of cartography can be divided into two general categories: general cartography and thematic cartography. General cartography involves those maps that are constructed for a general audience and thus contain a variety of features. General maps exhibit many reference and location systems and often are produced in a series. For example, the 1:24,000 scale topographic maps of the United States Geological Survey (USGS) are a standard as compared to the 1:50,000 scale Canadian maps. The government of the UK produces the classic 1:50,000 (replacing the older 1 inch to 1 mile) "Ordnance Survey" maps of the entire UK and with a range of correlated larger- and smaller-scale maps of great detail.

Thematic cartography involves maps of specific geographic themes, oriented toward specific audiences. A couple of examples might be a dot map showing corn production in Indiana or a shaded area map of Ohio counties, divided into numerical choropleth classes. As the volume of geographic data has exploded over the last century, thematic cartography has become increasingly useful and necessary to interpret spatial, cultural and social data. An orienteering map combines both general and thematic cartography, designed for a very specific user community. The most prominent thematic element is shading, that indicates degrees of difficulty of travel due to vegetation. The vegetation itself is not identified, merely classified by the difficulty ("fight") that it presents.

Topographic vs. Topological

A topographic map is primarily concerned with the topographic description of a place, including (especially in the 20th and 21st centuries) the use of contour lines showing elevation. Terrain or relief can be shown in a variety of ways.

A topological map is a very general type of map, the kind one might sketch on a napkin. It often disregards scale and detail in the interest of clarity of communicating specific route or relational information. Beck's London Underground map is an iconic example. Though the most widely used map of "The Tube," it preserves little of reality: it varies scale constantly and abruptly, it straightens curved tracks, and it contorts directions. The only topography on it is the River Thames, letting the reader know whether a station is north or south of the river. That and the topology of station order and interchanges between train lines are all that is left of the geographic space. Yet those are all a typical passenger wishes to know, so the map fulfils its purpose.

Map Purpose and Selection of Information

Arthur H. Robinson, an American cartographer influential in thematic cartography, stated that a map not properly designed "will be a cartographic failure." He also claimed, when considering all aspects of cartography, that "map design is perhaps the most complex." Robinson codified the mapmaker's understanding that a map must be designed foremost with consideration to the audience and its needs.

From the very beginning of mapmaking, maps "have been made for some particular purpose or set of purposes". The intent of the map should be illustrated in a manner in which the percipient acknowledges its purpose in a timely fashion. The term *percipient* refers to the person receiving information and was coined by Robinson. The principle of figure-ground refers to this notion of engaging the user by presenting a clear presentation, leaving no confusion concerning the purpose of the map. This will enhance the user's experience and keep his attention. If the user is unable to identify what is being demonstrated in a reasonable fashion, the map may be regarded as useless.

Making a meaningful map is the ultimate goal. Alan MacEachren explains that a well designed map "is convincing because it implies authenticity" (1994, pp. 9). An interesting map will no doubt engage a reader. Information richness or a map that is multivariate shows relationships within the map. Showing several variables allows comparison, which adds to the meaningfulness of the map. This also generates hypothesis and stimulates ideas and perhaps further research. In order to convey the message of the map, the creator must design it in a manner which will aid the reader in the overall understanding of its purpose. The title of a map may provide the "needed link" necessary for communicating that message, but the overall design of the map fosters the manner in which the reader interprets it (Monmonier, 1993, pp. 93).

In the 21st century it is possible to find a map of virtually anything from the inner workings of the human body to the virtual worlds of cyberspace. Therefore there are now a huge variety of different styles and types of map – for example, one area which has evolved a specific and recognisable variation are those used by public transport organisations to guide passengers, namely urban rail and metro maps, many of which are loosely based on 45 degree angles as originally perfected by Harry Beck and George Dow.

Naming Conventions

Most maps use text to label places and for such things as the map title, legend and other information. Although maps are often made in one specific language, place names often differ between languages. So a map made in English may use the name *Germany* for that country, while a German map would use *Deutschland* and a French map *Allemagne*. A non-native term for a place is referred to as an exonym.

In some cases the correct name is not clear. For example, the nation of Burma officially changed its name to Myanmar, but many nations do not recognise the ruling junta and continue to use *Burma*. Sometimes an official name change is resisted in other languages and the older name may remain in common use. Examples include the use of *Saigon* for Ho Chi Minh City, *Bangkok* for Krung Thep and *Ivory Coast* for Cτte d'Ivoire.

Difficulties arise when transliteration or transcription between writing systems is required. Some well-known places have well-established names in other languages and writing systems, such as *Russia* or *Rußland* for Ðîññèµÿ, but in other cases a system of transliteration or transcription is required. Even in the former case, the exclusive use of an exonym may be unhelpful for the map user. It will not be much use for an English user of a map of Italy to show Livorno *only* as "Leghorn" when road signs and railway timetables show it as "Livorno". In transliteration, the characters in one script are represented by characters in another. For example, the Cyrillic letter *Ð* is usually written as *R* in the Latin script, although in many cases it is not as simple as a one-for-one equivalence. Systems exist for transliteration of Arabic, but the results may vary. For example, the Yemeni city of Mocha is written variously in English as Mocha, Al Mukha, al-Mukhβ, Mocca and Moka. Transliteration systems are based on relating written symbols to one another, while transcription is the attempt to spell in one language the phonetic sounds of another. Chinese writing is now usually converted to the Latin alphabet through the Pinyin phonetic transcription systems. Other systems were used in the past, such as Wade-Giles, resulting in the city being spelled *Beijing* on newer English maps and *Peking* on older ones.

Further difficulties arise when countries, especially former colonies, do not have a strong national geographic naming standard. In such cases, cartographers may have to choose between various phonetic spellings of local names versus older imposed, sometimes resented, colonial names. Some countries have multiple official languages, resulting in multiple official placenames. For example, the capital of Belgium is both *Brussel* and *Bruxelles.* In Canada, English and French are official languages and places have names in both languages. British Columbia is also officially named *la Colombie-Britannique.* English maps rarely show the French names outside of Quebec, which itself is spelled *Québec* in French.

The study of placenames is called toponymy, while that of the origin and historical usage of placenames as words is etymology.

In order to improve legibility or to aid the illiterate, some maps have been produced using pictograms to represent places. The iconic example of this practice is Lance Wyman's early plans for the Mexico City Metro, on which stations were shown simply as stylized logos. Wyman also prototyped such a map for the Washington Metro, though ultimately the idea was rejected. Other cities experimenting with such maps are Fukuoka, Guadalajara and Monterrey.

Map Symbology

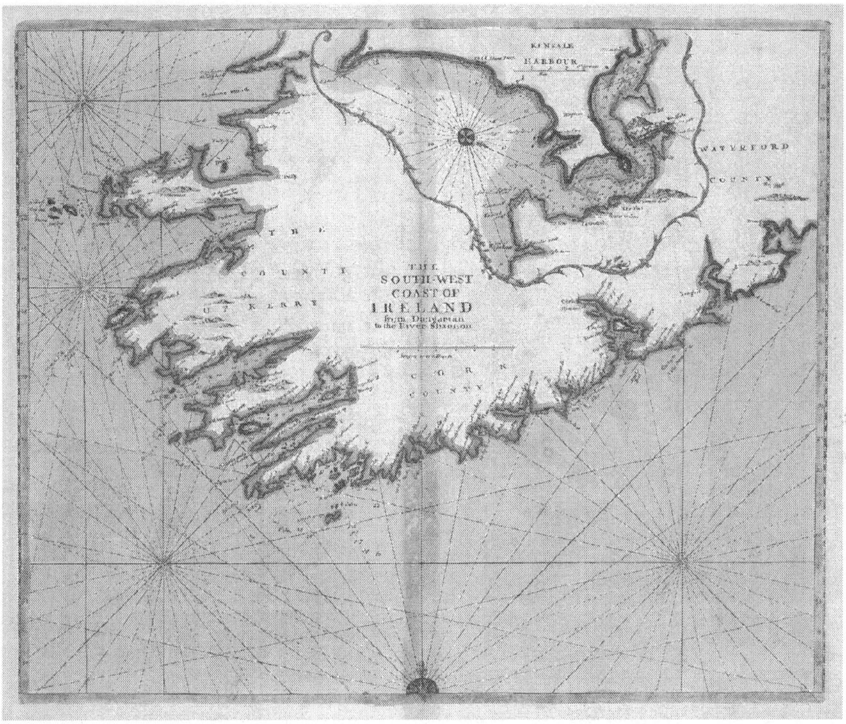

Figure: *A map of the southwest coast of Ireland created in the early 18th century. Notice the north arrow at the bottom of the map. Also, colours are used in the map to distinguish different geographical areas.*

The quality of a map's design affects its reader's ability to extract information and to learn from the map. Cartographic symbology has been developed in an effort to portray the world accurately and effectively convey information to the map reader. A legend explains the pictorial language of the map, known as its symbology. The title indicates the region the map portrays; the map image portrays the region and so on. Although every map element serves some purpose, convention only dictates inclusion of some elements, while others are considered optional. A menu of map elements includes the neatline (border), compass rose or north arrow, overview map, bar scale, map projection and information about the map sources, accuracy and publication.

When examining a landscape, scale can be intuited from trees, houses and cars. Not so with a map. Even such a simple thing as a north arrow is crucial. It may seem obvious that the top of a map should point north, but this might not be the case.

Map colouring is also very important. How the cartographer displays the data in different hues can greatly affect the understanding or feel of the map. Different intensities of hue portray different objectives the cartographer is attempting to get across to the audience. Today, personal computers can display up to 16 million distinct colours at a time. This fact allows for a multitude of colour options for even for the most demanding maps. Moreover, computers can easily hatch patterns in colours to give even more options. This is very beneficial, when symbolizing data in categories such as quintile and equal interval classifications. Quantitative symbols give a visual measure of the relative size/importance/number that a symbol represents and to symbolize this data on a map, there are two major classes of symbols used for portraying quantitative properties. Proportional symbols change their visual weight according to a quantitative property. These are appropriate for extensive statistics. Choropleth maps portray data collection areas, such as counties or census tracts, with colour. Using colour this way, the darkness and intensity (or value) of the colour is evaluated by the eye as a measure of intensity or concentration.

Map Generalization

A good map has to compromise between portraying the items of interest (or themes) in the right place on the map, and the need to show that item using text or a symbol, which take up space on the map and might displace some other item of information. The cartographer is thus constantly making judgements about what to include, what to leave out and what to show in a *slightly* incorrect place. This issue assumes more importance as the scale of the map gets smaller (i.e. the map shows a larger area) because the information shown on the map takes up more space *on the ground.* A good example from the late 1980s was the Ordnance Survey's first digital maps, where the *absolute* positions of major roads were sometimes a scale distance of hundreds of metres away from ground truth, when shown on digital maps at scales of 1:250,000 and 1:625,000, because of the overriding need to annotate the features.

Map Projections

The Earth being spherical, any flat representation generates distortions such that shapes and areas cannot both be conserved simultaneously, and distances can never all be preserved. The mapmaker must choose a suitable map projection according to the space to be mapped and the purpose of the map.

Cartographic Errors

Some maps contain deliberate errors or distortions, either as propaganda or as a "watermark" to help the copyright owner identify infringement if the error appears in competitors' maps. The latter often come in the form of nonexistent, misnamed, or misspelled "trap streets". Other names and forms for this are paper townsites, fictitious entries, and copyright easter eggs. Another motive for deliberate errors is cartographic "vandalism": a mapmaker wishing to leave his or her mark on the work. Mount Richard, for example, was a fictitious peak on the Rocky Mountains' continental divide that appeared on a Boulder County, Colorado map in the early 1970s. It is believed to be the work of draftsman Richard Ciacci. The fiction was not discovered until two years later. Sandy Island (New Caledonia) is an example of a fictitious location that stubbornly survives, reappearing on new maps copied from older maps while being deleted from other new editions. The time and reason for its original placement on maps is unknown.

Geovisualization

Geovisualization or Geovisualisation, short for *Geographic Visualization*, refers to a set of tools and techniques supporting geospatial data analysis through the use of interactive visualization.

Like the related fields of scientific visualization and information visualization geovisualization emphasizes knowledge construction over knowledge storage or information transmission. To do this, geovisualization communicates geospatial information in ways that, when combined with human understanding, allow for data exploration and decision-making processes.

Traditional, static maps have a limited exploratory capability; the graphical representations are inextricably linked to the geographical information beneath. GIS and geovisualization allow for more interactive maps; including the ability to explore different layers of the map, to zoom in or out, and to change the visual appearance of the map, usually on a computer display. Geovisualization represents a set of cartographic technologies and practices that take advantage of the ability of modern microprocessors to render changes to a map in real time, allowing users to adjust the mapped data on the fly.

The term visualization is first mentioned in the cartographic literature at least as early as 1953, in an article by University of Chicago geographer Allen K. Philbrick. New developments in the field of computer science prompted the National Science Foundation to redefine the term in a 1987 report which placed visualization at the convergence of computer graphics, image processing, computer vision, computer-aided design, signal processing, and user interface studies and emphasized both the knowledge creation and hypothesis generation aspects of scientific visualization.

Geovisualization developed as a field of research in the early 1980s, based largely on the work of French graphic theorist Jacques Bertin. Bertin's work on cartographic design and information visualization share with the National Science Foundation report a focus on the potential for the use of "dynamic visual displays as prompts for scientific insight and on the methods through which dynamic visual displays might leverage perceptual cognitive processes to facilitate scientific thinking".

Geovisualization has continued to grow as a subject of practice and research. The International Cartographic Association (ICA) established a Commission on Visualization & Virtual Environments in 1995.

Related Fields

Geovisualization is closely related to other visualization fields, such as scientific visualization and information visualization. Owing to its roots in cartography, geovisualization contributes to these other fields by way of the map metaphor, which "has been widely used to visualize non-geographic information in the domains of information visualization and domain knowledge visualization. It is also related to urban simulation.

Practical Applications

Geovisualization has made inroads in a diverse set of real-world situations calling for the decision-making and knowledge creation processes it can provide. The following list provides a summary of some of these applications as they are discussed in the geovisualization literature.

Wildland Fire Fighting

Firefighters have been using sandbox environments to rapidly and physically model topography and fire for wildfire incident command strategic planning. The SimTable is a 3D interactive fire simulator, bringing sandtable exercises to life. The SimTable uses advanced computer simulations to model fires in any area, including local neighbourhoods, utilizing actual slope, terrain, wind speed/direction, vegetation, and other factors. SimTable Models were used in Arizona's largest fire on record, the Wallow_Fire.

Forestry

Geovisualizers, working with European foresters, used CommonGIS and Visualization Toolkit (VTK) to visualize a large set of spatio-temporal data related to European forests, allowing the data to be explored by non-experts over the Internet. The report summarizing this effort "uncovers a range of fundamental issues relevant to the broad field of geovisualization and information visualization research".

The research team cited the two major problems as the inability of the geovisualizers to convince the foresters of the efficacy of geovisualization in their work and the foresters' misgivings over the dataset's accessibility to non-experts engaging in "uncontrolled exploration". While the geovisualizers focused on the ability of geovisualization to aid in knowledge construction, the foresters preferred the information-communication role of more traditional forms of cartographic representation.

Archaeology

Geovisualization provides archaeologists with a potential technique for mapping unearthed archaeological environments as well as for accessing and exploring archaeological data in three dimensions.

The implications of geovisualization for archaeology are not limited to advances in archaeological theory and exploration but also include the development of new, collaborative relationships between archaeologists and computer scientists.

Environmental Studies

Geovisualization tools provide multiple stakeholders with the ability to make balanced environmental decisions by taking into account the "the complex interacting factors that should be taken into account when studying environmental changes". Geovisualization users can use a georeferenced model to explore a complex set of environmental data, interrogating a number of scenarios or policy options to determine a best fit.

Urban Planning

Both planners and the general public can use geovisualization to explore real-world environments and model 'what if' scenarios based on spatio-temporal data. While geovisualization in the preceding fields may be divided into two separate domains—the private domain, in which professionals use geovisualization to explore data and generate hypotheses, and the public domain, in which these professionals present their "visual thinking" to the general public—planning relies more heavily than many other fields on collaboration between the general public and professionals.

Planners use geovisualization as a tool for modelling the environmental interests and policy concerns of the general public. Jiang et al. mention two examples, in which "3D photorealistic representations are used to show urban redevelopment [and] dynamic computer simulations are used to show possible pollution diffusion over the next few years." The widespread use of the Internet by the general public has implications for these collaborative planning efforts, leading to increased participation by the public while decreasing the amount of time it takes to debate more controversial planning decisions.

City Map

A city map is a large-scale thematic map of a city (or part of a city) created to enable the fastest possible orientation in an urban space. The graphic representation of objects on a city map is therefore usually greatly simplified, and reduced to generally understood *symbology*.

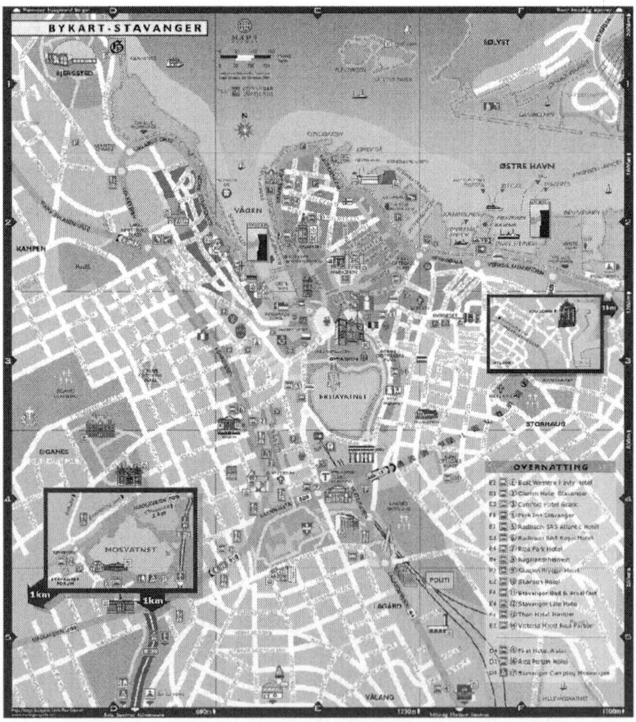

Figure: *City map of Stavanger (Norway).*

Depending upon its target group or market, a city map will include not only the city's transport network, but also other important information, such as city sights or public institutions.

Content and Design

The scale of a city map is usually between 1:10,000 and 1:25,000. Densely settled downtown areas will sometimes be partly drawn in a larger scale, on a separate detail map.

In addition to linear true to scale maps, there are also maps with variable scale, for example where the scale gradually increases towards the city centre (a fish-eye projection).

Central to the information provided by a city map is the street network, including its street names (often supplemented by at least a selection of individual house numbers), along with buildings, parks and waterways. Streets and points of interest are usually also listed in a legend or register, locating objects on a map grid on the map. Important places such as administrative buildings, cultural institutions, attractions, etc. may be highlighted with the assistance of pictograms. The map may also be complemented by representations of public transport facilities.

History

As early as the time of the Ancient Near East, clay tablets were being produced with scaled, graphical representations of cities. Excavations of the Sumerian city of Nippur brought to light a fragment of an approximately 3,500-year-old city map, which is sometimes referred to as the oldest known city map. The clay tablet depicts the temple of Enlil, a city park, the city wall including its gates, along with a canal and the river Euphrates. The individual objects on this map were already labelled, in a Sumerian cuneiform.

Late Middle Ages

In manuscripts and early printed books of the Late Middle Ages, cities are often shown in profile, or viewed from an elevated standpoint. Nautical charts of that time sometimes depict partly stylized cityscapes drawn in pictogram form - for example in Cristoforo Buondelmonti's *Liber insularum archipelagi* (Book of Islands), from the year 1422.

The Nuremberg Chronicle, which first appeared in 1493, is one of the most important collections of city views of the late Middle Ages, with over 100 such illustrations. Nevertheless, panoramas like this one, or the one in Bernhard von Breydenbach's *Travelogue* from 1483, had more narrative or representative functions.

Illustrated in these works are the local conditions and relevant characteristics - such as ports, magnificent buildings, walls, etc. - as a background for highlighting historical descriptions or economic benefits of the city. On the other hand, less emphasis was placed on accuracy: in the Nuremberg Chronicle, merely one quarter of the city views represented the actual appearance of the subject city, and some individual images were even used simultaneously to represent several different cities.

Renaissance

In the 16th century, the artists and scholars of the Renaissance had extensive knowledge of mathematical perspectives and projections. This knowledge also affected the work of cartographers and the production of cityscapes (especially in Italy). A key innovation was that the city was no longer portrayed simply from an imaginary or real perspective, but drawn initially as a two-dimensional map, and then, using a process of accurate perspective drawing, transformed into a three-dimensional image. An early example of a geometrically exact and highly detailed work of this kind is the city map of Venice created by Jacopo de' Barbari in around 1500.

Whereas the illustrations of the late Middle Ages are usually still simple small-format woodcuts, an increasingly common process from 1500 onwards was the creation of prints from huge woodcuts and woodblocks. Jacopo de' Barbari's map of Venice was already as large as 139 centimetres (55 in) x 282 centimetres (111 in), and consisted of six individual panels. From the middle of the 16th century, the copperplate process, originating in Antwerp, began to compete with the woodcut, and allowed far more refined and detailed illustrations.

19th Century Pocket Atlasses

One of the first City pocket atlasses, and the first pocket atlas of London, was "Collins' Illustrated Atlas of London" published in 1854 and drawn and engraved by Richard Jarman.

Road Map

A road map or route map is a map that primarily displays roads and transport links rather than natural geographical information. It is a type of navigational map that commonly includes political boundaries and labels, making it also a type of political map. In addition to roads and boundaries, road maps often include points of interest, such as prominent businesses or buildings, tourism sites, parks and recreational facilities, hotels and restaurants, as well as airports and train stations. A road map may also document non-automotive transit routes, although often these are found only on transit maps.

The Turin Papyrus Map is sometimes characterized as the earliest known road map. Drawn around 1160 BC, it depicts routes along dry river beds through a mining region east of Thebes in Ancient Egypt. The Dura-Europos Route map is the oldest known map of (a part of) Europe preserved in its original form. It is a fragment of a map drawn onto a leather portion of a shield by a Roman soldier in c. 235 AD. It depicts several towns along the northwest coast of the Black Sea.

The Tabula Peutingeriana, a copy of a scroll originally dating to about 350 AD, plots the extent of the *Cursus publicus*, the Roman road network that ran from Europe and North Africa to West Asia. It is highly schematic, compressing the Mediterranean Sea to a sliver and orienting the Italian Peninsula to run east-west.

The Gough Map, dating to about 1360, is the oldest known road map of Great Britain.

In 1500, Erhard Etzlaub produced the "Rom-Weg" (*Way to Rome*) Map, the first known road map of medieval Central Europe. It was produced to help religious pilgrims reach Rome for the occasion of the "Holy Year 1500".

Rand McNally's first road map, the *New Automobile Road Map of New York City & Vicinity*, was published in 1904. Gousha was founded in 1926 by former Rand McNally employees. General Drafting was founded in 1909. These three companies produced most of the approximately eight billion free maps handed out at American filling stations over a period of about 1920 to 1980. The practice of offering free maps diminished considerably in the 1970s.

The first Michelin map was produced in 1910.

With the rise of GPS navigation and other electronic maps in the 21st century, the use of printed maps is waning.

Itineraria

An alternative to, and in many ways the precursor of the road map, was the *itinerarium*, a listing of towns and other stops, with intervening distances. The Tabula Peutingeriana, mentioned above, is in effect an itinerarium in visual form, offering routes and distances with little geographical accuracy.

Types

Road maps come in many shapes, sizes and scales. Small, single-page maps may be used to give an overview of a region's major routes and features. Folded maps can offer greater detail covering a large region. Electronic maps typically present a dynamically generated display of a region, with its scale, features, and level of detail specified by the user.

Road maps can also vary in complexity, from a simple schematic map used to show how to get to a single specific destination (such as a business), to a complex electronic map, which may layer together many different types of maps and information – such as a road map plotted over a topographical 3D satellite image (a viewing mode frequently used within Google Earth). Highway maps generally give an overview of major routes within a medium to large region ranging from a few dozen to a few thousand miles or kilometres.

Street maps usually cover an area of a few miles or kilometres (at most) within a single city or extended metropolitan area. City maps are generally a specialised form of street map.

A road atlas is a collection of road maps covering a region as small as a city or as large as a continent, typically bound together in a book. Spiral binding is a popular format for road atlases, to permit lay-flat usage and to reduce wear and tear. Atlases may cover a number of discrete regions, such as all of the states or provinces of a given nation, or a single continuous region in high detail split across several pages.

Many motoring organisations, especially those in the European Union, North America, Australia and New Zealand produce road maps.

Common Features

Road maps often distinguish between major and minor thoroughfares (such as motorways vs. surface streets) by using thicker lines or bolder colours for the major roads. Printed road maps commonly include an index of cities and other destinations found on the map; smaller-scale maps often include indexes of streets and other routes. These indexes give the location of the feature on the map via a grid reference. Inset maps may be used to provide greater detail for a specific area, such as a city map inset into a map of a state or province.

Graphic Variables

Maps use graphic symbols to represent the locations and attributes of phenomena distributed across the Earth's surface. Variations in symbol size, colour lightness, colour hue, and shape can be used to represent quantitative and qualitative variations in attribute data. By convention, each of these "graphic variables" is used to represent a particular type of attribute data.

Purpose and Development

Our usual way of communicating is with words. Written words consist of single symbols (letters), gaining meaning when arranged in certain combinations. The question is: If there are basic visual symbols arranged in a particular way, can they be used to convey information in a similar manner? All of those developments were primarily made for cartographic purposes. With the computerization of information these visual variables were adapted and used for information visualization. The concept of information visualization began in the 1930's, but after the 1950's became more developed by cartographers. Since the development of computers has revolutionized all aspects of information visualization. Mapmaking with pen and paper became unnecessary. Map making software such as CAD, GIS and specialised map illustration software became very important.

Basics

Maps are a form of visual communication. You learned about map communication in the second module. To effectively communicate with maps requires that the map maker and map user have a similar understanding of the cartographic language used to encode the map information.

This process is similar to sending a telegraph. The sender's words in a telegraph must be converted into code which is sent over a channel to another location. There, the code must be decoded into the words of the message. The message received at the end of this process should be the same as the one the person sending the telegraph wished to send. However, if the different code is used at the two ends of the channel, the message will be incomprehensible. Luckily, telegraph operators use standardized codes, so mistakes are not often made. Like you see in the following graphics, a map acts as a channel between its producer and user. Hopefully, they use the same code, too.

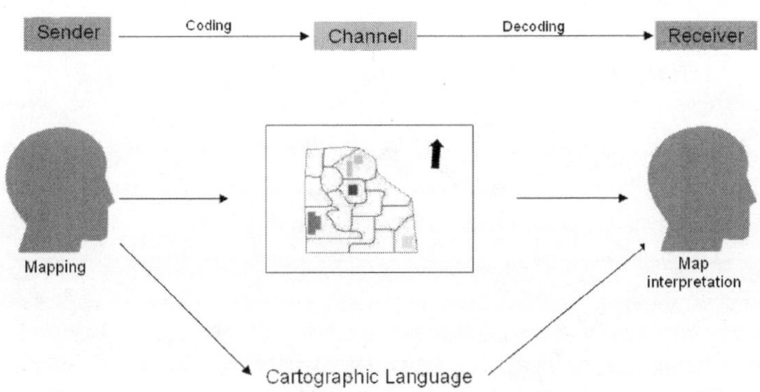

In map making, the same content can be represented many different ways, using many different codes. There are no absolute rules about how to represent information. There are, however, useful heuristics and some good symbolization theories. Common sense plays a large role as well.

Unfortunately, there are numerous examples of how to not to make a legible map. To create a map that is clear and intuitive to others requires careful consideration of each element. The choice of map symbols depends on the map scale and the data level of the phenomena to be displayed.

Visual Variables

Jacques Bertin, a French cartographer and information designer described one of the first comprehensive theories on graphic variables and their uses. In addition to location, he named six visual variables that can be used to differentiate objects in a map. Others built on his work and expanded the list to include: location in space (x-, y-, z-coordinates), size, shape, colour value, hue, saturation, orientation, texture, arrangement and focus. These graphic variables are the building blocks for mapmaking.

Use of these variables is also limited by the ability of the human eye to see differences. Each variable has a minimum spread and a maximum length. Spread is the ratio of the smallest to the largest value, for example the ratio of areas or line thicknesses. For circles with a diameter of 0.3 mm to 3 cm the spread is 1:10,000. Size differences must be sufficiently large (low ratio) for map users to identify high and low values. Length is the number of possible values for each variable. For example, lines displayed at 0.2 mm thickness can typically be divided into no more than five gray values.

Not all of the variables are appropriate for every kind of data. You learned about scale levels in module 5. Maps for non-experts typically display nominal or ordinal data, as extracting numerical information from a map is difficult. Size, colour value, and colour saturation are best suited for ordered data and can be applied to numerical differentiation. Bertin argued that only size and location are truly appropriate variables for depicting quantities.

Size

Size is the only visual variable that works for all quantitative attributes. The size of the displayed object reflects the unit value of the theme of interest. Size is limited by the ability of the map reader to estimate the difference between sizes. Size differences may be displayed by adjusting linearly or by adjusting the area, for example, by doubling the length of the sides of a square or by doubling its area (the sides increase by about 40%). Experiments have shown no clear answer as to which is the better way. Each method results in estimation errors from 10-30%. It is recommended to use a ratio close to the obiect area.

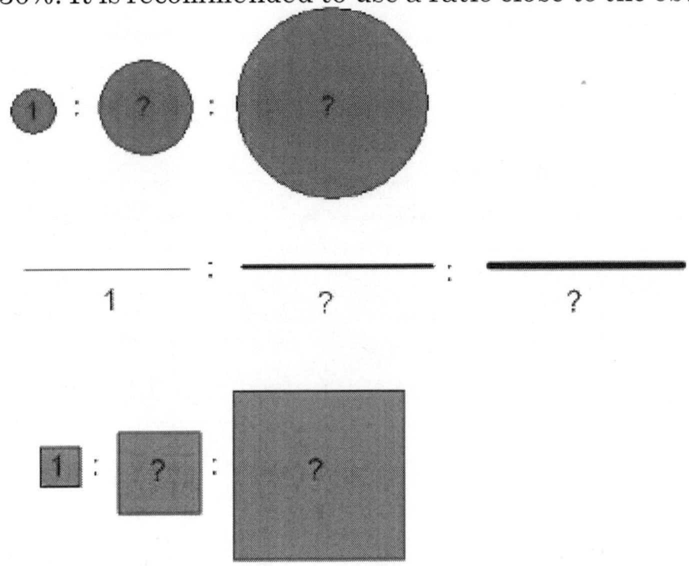

Value

Three attributes of colour, value, hue and saturation, can be used as graphic variables. Colour value is the lightness or darkness of a colour from pure black to pure white. High values are light (for example white or yellow) and low values are dark (for example black or navy blue). Value is often used for ordered attributes and has limited application to quantitative attributes, because visual perception of value is not linear. As noted before, the human ability to differentiate shades of gray is limited. In printing, value is changed according to the amount of ink per area, or a rasterization method. Varying colour values combined with hue variations results in a higher contrast map than if value is applied alone. This method aids in displaying ordinal data.

Hue (Colour)

For most people hue is synonymous with colour. It is the measure of wavelengths that a surface reflects resulting in perceived differences in red, green, blue, etc.

Visible light reflects in wavelengths from 380 nm (violet) to 760 nm (red). The human eye is well developed to perceive colours, with the exception of the about 8% of people with inherited colour-blindness (we will discuss more in module 8). We enjoy looking at colourful images, and this tendency leads often to overuse of colour in maps.

Variations in colour hue are excellent for nominal attributes and can also be used metaphorically, for example with administrative units, temperatures, danger levels. Hue alone has limited use for ordered and quantitative attributes.

Colour hues do have a natural order, according to their wavelength, but many people are not familiar with the Red-Orange-Yellow-Green-Blue-Voilet spectrum. Using this spectrum to display ordinal data from red to violet results in confusion. Colour hue variations appear also as value differences. Yellow appears lighter than other hues, so applying the full colour scale makes it appear that midrange data is lower than it really is. Using half of the scale, from yellow to red or from yellow to blue or violet, combines the attributes of hue and value to create a more eye-catching map than a simple gray scale.

We will explore colour more in module 8, but for now you should keep in mind these basic rules: areas in light, lines in dark colours, abide by accepted colour conventions (for example water=blue). Pay attention to colour harmonies.

Saturation

The third dimension of colour is saturation, which denotes how pure the colour is compared to a gray tone of the same value.

Shape

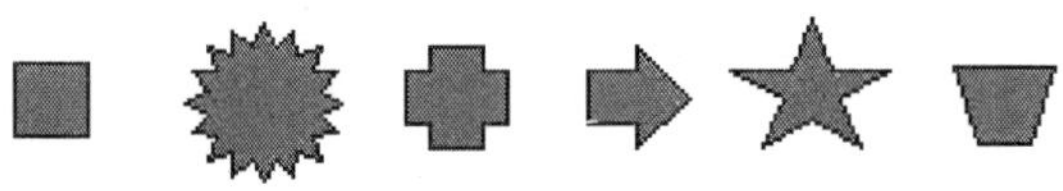

Shape is used to denote nominal attributes, most often at points. Similar to colour, there are certain conventions on shape use. Shapes are not as easy to recognise and differentiate at a glance as colour and the object-referent relationship may not be as strong. Shapes vary from the abstract to the ideographic. The more similar a shape is to its referent, the easier it is for map users to remember what the shape represents. Abstract shapes often require the map user to refer to the legend repeatedly. Symbol recognition also diminishes with decreasing shape size and increasing number of shapes. You may have seen natural resource maps that illustrate this problem. The distribution pattern of individual resources is difficult to recognise in the overview map. A map with one or two symbols is easier to interpret.

Orientation

Orientation is used for cyclic quantitative attributes at points. It is best suited for vector fields, such as wind, gravity and magnetism.

The map symbols used by meteorologists to denote wind direction and speed combine orientation with shape. A maximum of four (or eight directed) orientations can be directly recognised.

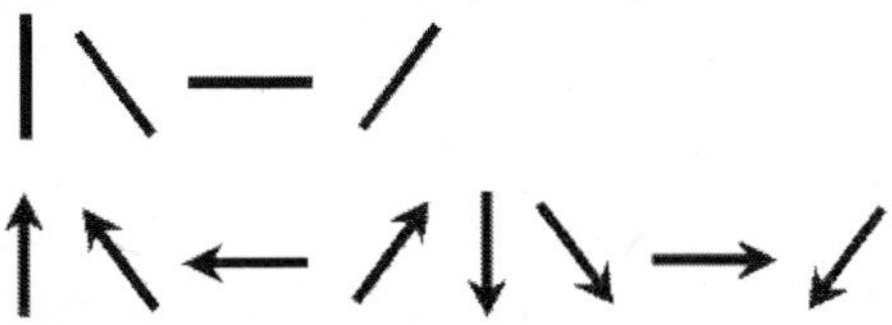

Texture

Texture or pattern is a complex graphic variable, combining size, value, hue, shape and orientation. Defined narrowly, texture or spacing refers to the spatial frequency of components of a pattern. It is used for nominal and ordered attributes of areas and lines. Texture has an almost unlimited length and can be used effectively to denote multiple phenomena in one map that would be confusing to represent with any of the other variables alone.

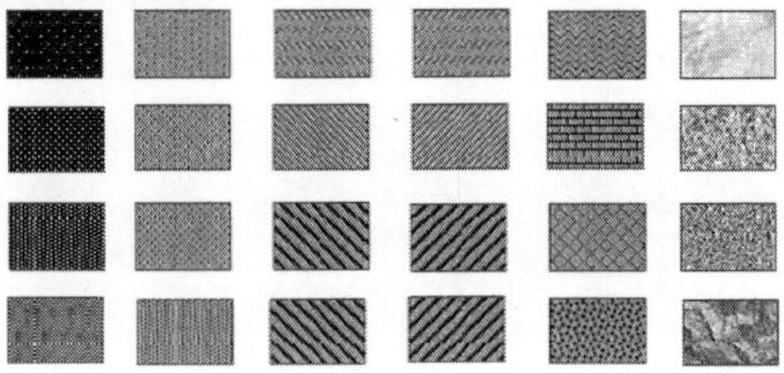

Jaques Bertin

Jaques Bertin described marks as these basic units and also developed a given number of methods through which these units can be modified, including position, size, shape, or colour. These predefined modifications are called visual variables. Each of these variables can have certain characteristics. Sometimes visual variables are also called visual attributes.

Marks

A mark is made to represent some information other than itself. It is also referred to as a sign.

Marks can be

- Points are dimensionless locations on the plane, represented by signs that obviously need to have some size, shape or colour for visualization.
- Lines represent information with a certain length, but no area and therefore no width. Again lines are visualized by signs of some thickness.

- Areas have a length and a width and therefore a two-dimensional size.
- Surfaces are areas in a three-dimensional space, but with no thickness.
- Volumes have a length, a width and a depth. They are thus truly three-dimensional.

Visual Variables: Jaques Bertin defined seven Visual Variables consisting of:

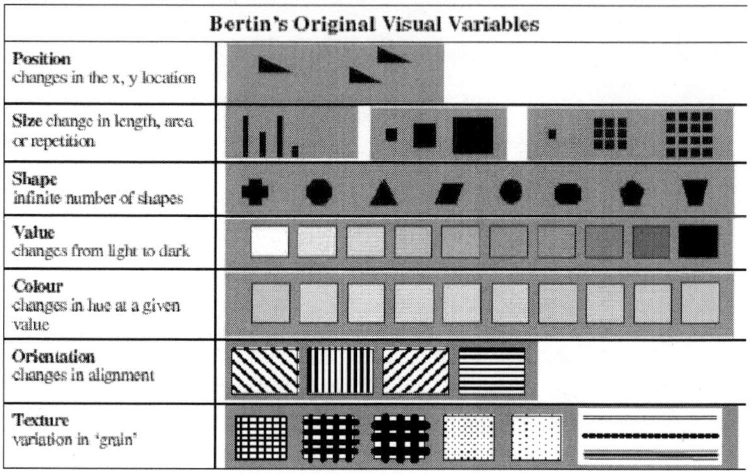

Jock D. Mackinley

Jock D. Mackinlay invented a number of Information Visualization techniques such as the information visualization reference model.

Visual Variables: The list of visual variables was later expanded by Jock D. Mackinlay. He also provided different sorting for their accuracy, based on the task.

Ranking of perceptual tasks:

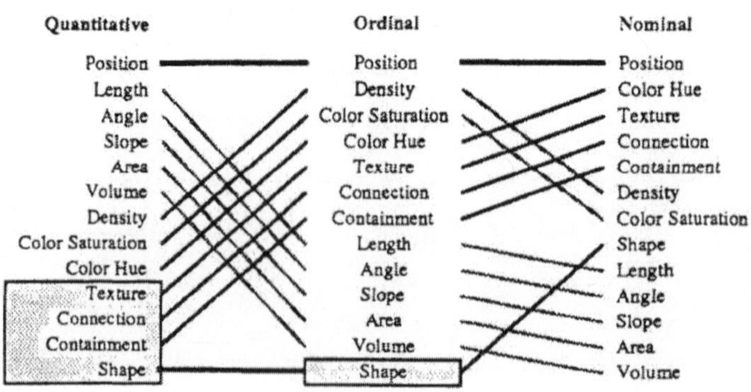

Recent Work

The list was further expanded by several later publications. Most of them are also grouping the visual variables, e.g. combining *length*, *area* and *repetition* to *shape* or breaking down *position* in the three dimensions of space and one time dimension.

Since nowadays information is presented by computers, the addition of motion as a new visual variable becomes important. Changes in motion can include direction, speed, frequency, rhythm, flicker, trails, and style.

Characteristics

The choice of the variable, which would be most appropriate to represent each aspect of information depends on its characteristics.

- Selective: If a mark changes in this variable and as an effect can be selected from the other marks easily the visual variable is said to be selective.
- Associative: Several marks can be grouped across changes in other visual variables.
- Quantitative: If the difference between two marks in this variable can be interpreted numerically, the visual variable is quantitative.
- Order: If the variable supports ordered reading it is an ordered visual variable. This means that a change could be read as more or less (e.g. in size you can order marks according to their area).
- Length: The length defines how many values the variable features. For example how many shades of grey can be recognised?

Usage

The process of mapping data to visual variables is called visual mapping. Choosing different visual variables for representing different aspects of the same information can greatly influence the perception and understanding of the presented information. It is therefore important to know and appropriately use the characteristics of visual variables when creating *any* visual data representation.

Perceptual and Cognitive Limitations

- symbol differences must be perceptible to be of use
  - JND - just noticeable difference - the smallest difference which can be reliably perceived between symbols, sizes, colours, shapes etc.

    - LPD - least practical difference - the smallest difference which can be produced by the cartographic process
- eye's sensitivity to various graphic codes
    - some codes "get through" better
        - e.g. use of yellow for fire trucks allows them to stand out better in the visual field
    - sensitivity varies across visual field
        - "peripheral vision" is enhanced by movement, varies among individuals
- cognitive aspects
    - indications that perception is dependent on cognition - knowledge understanding of phenomena
    - colour categories/nameability - certain colours may have associations with names, concepts

E. Graphic Limits

- digital devices provide finite resolution
- spatial - where symbols might be and their shapes
    - display device has a set screen or paper size
    - display pixels have a set size, finite number of spatial locations
    - aliasing - line (or point) mapped onto closest pixel(s)
        - produces stepped (straight) lines
- colour - what colours things might be
    - limit on number of colours available (palette) - plotter may have only 8 pen colours - screen may have millions of possible colours
    - limit on range of luminance and contrast
    - how many colours displayable at one time - 2n where n is the number of bit planes
    - what the colours are
- temporal limits
    - data retrieval from mass storage or from core memory?
    - how much data processing needed to compute display?
    - writing to the display device
        - speed limited by communication overhead & bus contention (competition from other activities)

- o these factors may preclude using some types of display image
  - – animation requires fast through-put
  - – complex images require fast data retrieval
- o acceptable response time
  - – people don't like to sense a pause in the system
  - – typical goal: maximum of two seconds for complex operations, instantaneous for all others
- o how long should something remain visible to be noticed?

F. Representing Uncertainty

- have to use SOME graphic code
  - o don't want its meaning confused with something else
    - – e.g. line drawn wide to represent uncertain position confused with wide highway or braided stream

Explicit uncertainty codes

- mark things which are uncertain with a colour
  - o e.g. red or yellow to suggest caution in using the information

Graphic ambiguity

- use graphic ambiguity to create cognitive/visual ambiguity
  - o e.g. multiple positions for an uncertainly located item
- dot density or colour could be used to show varying probability, e.g. a cloud with highest density in the centre
- absence of "hard" lines or edges where they are uncertain

Examples

- show uncertain area with a red tint overlay
- show uncertain lines as multiple lines (like a braided stream)
- fuzzy line
  - o vary the value or saturation of the line across its width
- blending between adjacent areas to show zones of transition
  - o blend the colours
  - o choose such that the blend works psychologically
    - – red &LT-> purple &LT-> blue
    - – blue &LT-> aqua &LT-> green
    - – NOT red &LT-> yellow &LT-> orange
  - o large set of possible colours are needed to show the appearance of a smooth transition

  - o transition can be simulated with a small set of colours by spatially blending pixel colours ("dithering")

G. Temporal Dependence

Basic strategies

- static maps
  - o show a single slice of time
  - o show several states at once by careful choice of symbols
  - o indicate amount or rate of change
- dynamic maps
  - o real time is compressed or scaled into changing display
  - o non-moving occurrences - events added and deleted at places through time
  - o moving objects - movement is animated on the screen - symbol is deleted at one location, regenerated at adjacent location

H. Showing a Third Dimension

Contours

- calculated contours (calculated by contouring algorithms)
  - o starting with a grid of elevations, thread contours and display the lines
- visual contours with elevation grid cells (contours are perceived but not computed explicitly)
  - o given a sufficiently dense raster of elevations
  - o shade pixels according to the elevation value of the central point using specified elevation ranges
- result is apparently (not analytically) a contour map

Hypsometric mapping

- set each pixel to a colour dependent on its height
- this is easily implemented as table look-up
- range of colours is conventional - dark green for low elevations, through green, yellow, brown, then white at highest elevations

Simulating oblique views of surface

- each pixel's illumination computed from its slope relative to simulated "sun"
  - o sun must be placed at top of image for correct visual perception - if sun is at bottom, eye sees surface inverted

- requires assumptions about reflectance of surface
  - o lakes, ice, some building materials produce highlights
- single light source makes the surface too "stark"
- assume light source infinitely far away from surface
  - o may assume viewer is also infinitely far away to avoid complex perspective calculations
- with TINs or coarse grids, edges of plane patches may be visible because of sharp change of slope
  - o discontinuities can be eliminated by varying intensity of illumination continuously over facets
  - o many 3-D display systems supply this capability - called Gouraud rendering

Local Analysis

The terms location and place in geography are used to identify a point or an area on the Earth's surface or elsewhere. The term *location* generally implies a higher degree of certainty than *place*, which often indicates an entity with an ambiguous boundary, relying more on human/social attributes of place identity and sense of place than on geometry.

Types of Location/Place

- A relative location is described as a displacement from another site—for example, "3 miles northwest of Seattle".
- A locality is likely to have a well-defined name but a boundary which is less well defined and which varies by context. London, for instance, has a legal boundary, but this is unlikely to completely match with general usage. An area within a town, such as Covent Garden in London, also almost always has some ambiguity as to its extent.

Absolute location - An absolute location is designated using a specific pairing of latitude and longitude in a Cartesian coordinate grid—for example, a Spherical coordinate system or an ellipsoid-based system such as the World Geodetic System—or similar methods.

An example would be the longitude and latitude of a place. For instance, the position of Lake Maracaibo in Venezuela can be expressed approximately in the WGS84 coordinate system as the location 9.80°N (latitude), 71.56°W (longitude). It is, however, important to remember that this is just one way the alternative ways can be seen in this link: 9.80°N 71.56°W.

Absolute location, however, is a term which has little real meaning, since any location must be expressed relative to something else. For example, longitude is the number of degrees east or west of the Prime Meridian, a line which has arbitrarily been chosen to pass through Greenwich, London. Similarly, latitude is the number of degrees north or south of the Equator. Because latitude and longitude are expressed *relative* to these lines, a position expressed in latitude and longitude is actually a relative location.

Place Identity

Place identity or place-based identity refers to a cluster of ideas about place and identity in the fields of geography, urban planning, urban design, landscape architecture, environmental psychology, ecocriticism and urban sociology/ecological sociology. It concerns the meaning and significance of places for their inhabitants and users. Methodologies for understanding place identity primarily involve qualitative techniques, such as interviewing, participant observation, discourse analysis and mapping a range of physical elements. Some urban planners, urban designers and landscape architects use forms of deliberative planning, design charettes and participatory design with local communities as a way of working with place identity to transform existing places as well as create new ones. This kind of planning and design process is sometimes referred to as placemaking.

Place identity is sometimes called urban character, neighbourhood character or local character.

Place identity has become a significant issue in the last 25 years in urban planning and design. Related to the worldwide movement to protect places with heritage significance, concerns have arisen about the loss of individuality and distinctiveness between different places as an effect of cultural globalisation.

Geographic Coordinate System

A geographic coordinate system is a coordinate system that enables every location on the Earth to be specified by a set of numbers or letters. The coordinates are often chosen such that one of the numbers represents vertical position, and two or three of the numbers represent horizontal position. A common choice of coordinates is latitude, longitude and elevation.

Geographic Latitude and Longitude

The "latitude" (abbreviation: Lat., φ, or phi) of a point on the Earth's surface is the angle between the equatorial plane and the straight line that passes through that point and is normal to the surface of a reference ellipsoid which approximates the shape of the Earth. This line passes a few kilometres away from the centre of the Earth except at the poles and the equator where it passes through Earth's centre. Lines joining points of the same latitude trace circles on the surface of the Earth called parallels, as they are parallel to the equator and to each other. The north pole is 90° N; the south pole is 90° S. The 0° parallel of latitude is designated the equator, the fundamental plane of all geographic coordinate systems. The equator divides the globe into Northern and Southern Hemispheres.

The "longitude" (abbreviation: Long., λ, or lambda) of a point on the Earth's surface is the angle east or west from a reference meridian to another meridian that passes through that point. All meridians are halves of great ellipses (often improperly called great circles), which converge at the north and south poles.

A line, which was intended to pass through the Royal Observatory, Greenwich (a suburb of London, UK), was chosen as the international zero-longitude reference line, the Prime Meridian. Places to the east are in the eastern hemisphere, and places to the west are in the western hemisphere. The antipodal meridian of Greenwich is both 180°W and 180°E. The zero/zero point is located in the Gulf of Guinea about 625 km south of Tema, Ghana.

In 1884 the United States hosted the International Meridian Conference and twenty-five nations attended. Twenty-two of them agreed to adopt the location of Greenwich as the zero-reference line. The Dominican Republic voted against the adoption of that motion, while France and Brazil abstained. To date, there exist organisations around the world which continue to use historical prime meridians which existed before the acceptance of Greenwich became commonplace.

The combination of these two components specifies the position of any location on the planet, but does not consider altitude nor depth. This latitude/longitude "webbing" is known as the *graticule*. A graticule representing latitude and longitude of the Earth does not constitute a hierarchy of geographical areas. This is to say, it is not an arrangement of related information or data.

UTM and UPS Systems

The Universal Transverse Mercator (UTM) and Universal Polar Stereographic (UPS) coordinate systems both use a metric-based cartesian grid laid out on a conformally projected surface to locate positions on the surface of the Earth. The UTM system is not a single map projection but a series of map projections, one for each of sixty 6-degree bands of longitude. The UPS system is used for the polar regions, which are not covered by the UTM system.

Stereographic Coordinate System

During medieval times, the stereographic coordinate system was used for navigation purposes. The stereographic coordinate system was superseded by the latitude-longitude system. Although no longer used in navigation, the stereographic coordinate system is still used in modern times to describe crystallographic orientations in the fields of crystallography, mineralogy and materials science.

Geodetic Height

To completely specify a location of a topographical feature on, in, or above the Earth, one has to also specify the vertical distance from the centre of the Earth, or from the surface of the Earth. Because of the ambiguity of "surface" and "vertical", it is more commonly expressed relative to a precisely defined vertical datum which holds fixed some known point. Each country has defined its own datum. For example, in the United Kingdom the reference point is Newlyn, while in Canada, Mexico and the United States, the point is near Rimouski, Quebec, Canada. The distance to Earth's centre can be used both for very deep positions and for positions in space.

Cartesian Coordinates

Every point that is expressed in ellipsoidal coordinates can be expressed as an x y z (Cartesian) coordinate. Cartesian coordinates simplify many mathematical calculations. The origin is usually the centre of mass of the earth, a point close to the Earth's centre of figure.

With the origin at the centre of the ellipsoid, the conventional setup is the expected right-hand:

Z-axis along the axis of the ellipsoid, positive northward

X- and Y-axis in the plane of the equator, X-axis positive toward 0 degrees longitude and Y-axis positive toward 90 degrees east longitude.

An example is the NGS data for a brass disk near Donner Summit, in California. Given the dimensions of the ellipsoid, the conversion from lat/lon/height-above-ellipsoid coordinates to X-Y-Z is straightforward—calculate the X-Y-Z for the given lat-lon on the surface of the ellipsoid and add the X-Y-Z vector that is perpendicular to the ellipsoid there and has length equal to the point's height above the ellipsoid. The reverse conversion is harder: given X-Y-Z we can immediately get longitude, but no closed formula for latitude and height exists." Using Bowring's formula in 1976 *Survey Review* the first iteration gives latitude correct within 10^{-11} degree as long as the point is within 10000 metres above or 5000 metres below the ellipsoid.

Shape of the Earth

The Earth is not a sphere, but an irregular shape approximating a biaxial ellipsoid. It is nearly spherical, but has an equatorial bulge making the radius at the equator about 0.3% larger than the radius measured through the poles. The shorter axis approximately coincides with axis of rotation. Map-makers choose the true ellipsoid that best fits their need for the area they are mapping. They then choose the most appropriate mapping of the spherical coordinate system onto that ellipsoid. In the United Kingdom there are three common latitude, longitude, height systems in use. The system used by GPS, WGS84, differs at Greenwich from the one used on published maps OSGB36 by approximately 112m. The military system ED50, used by NATO, differs by about 120m to 180m.

Though early navigators thought of the sea as a flat surface that could be used as a vertical datum, this is not actually the case. The Earth has a series of layers of equal potential energy within its gravitational field. Height is a measurement at right angles to this surface, roughly toward the centre of the Earth, but local variations make the equipotential layers irregular (though roughly ellipsoidal). The choice of which layer to use for defining height is arbitrary. The reference height that has been chosen is the one closest to the average height of the world's oceans. This is called the geoid.

The Earth is not static as points move relative to each other due to continental plate motion, subsidence, and diurnal movement caused by the Moon and the tides. The daily movement can be as much as a metre. Continental movement can be up to 10 cm a year, or 10 m in a century. A weather system high-pressure area can cause a sinking of 5 mm. Scandinavia is rising by 1 cm a year as a result of the melting of the ice sheets of the last ice age, but neighbouring Scotland is rising by only 0.2 cm. These changes are insignificant if a local datum is used, but are statistically significant if the global GPS datum is used.

Expressing Latitude and Longitude as Linear Units

On the GRS80 or WGS84 spheroid at sea level at the equator, one latitudinal second measures *30.715 metres*, one latitudinal minute is *1843 metres* and one latitudinal degree is *110.6 kilometres*. The circles of longitude, meridians, meet at the geographical poles, with the west-east width of a second naturally decreasing as latitude increases. On the equator at sea level, one longitudinal second measures *30.92 metres*, a longitudinal minute is *1855 metres* and a longitudinal degree is *111.3 kilometres*. At 30° a longitudinal second is *26.76 metres*, at Greenwich (51° 28' 38" N) *19.22 metres*, and at 60° it is *15.42 metres*.

On the WGS84 spheroid, the length in metres of a degree of latitude at latitude φ (that is, the distance along a north-south line from latitude (φ - 0.5) degrees to (φ + 0.5) degrees) is about

$$111132.954 - 559.822\cos 2\varphi + 1.175\cos 4\varphi$$

(Those coefficients can be improved, but as they stand the distance they give is correct within a centimetre.)

To estimate the length of a longitudinal degree at latitude ϕ we can assume a spherical Earth (to get the width per minute and second, divide by 60 and 3600, respectively):

$$\frac{\pi}{180} M_r \cos\phi$$

where Earth's average meridional radius M_r is 6,367,449 m. Since the Earth is not spherical that result can be off by several tenths of a percent; a better approximation of a longitudinal degree at latitude is

$$\frac{\pi}{180} a \cos\beta$$

where Earth's equatorial radius a equals *6,378,137 m* and ; $\tan\beta = \frac{b}{a}\tan\phi$ for the GRS80 and WGS84 spheroids, b/a calculates to be 0.99664719. (is known as the reduced (or parametric) latitude). Aside from rounding, this is the exact distance along a parallel of latitude; getting the distance along the shortest route will be more work, but those two distances are always within 0.6 metre of each other if the two points are one degree of longitude apart.

Longitudinal Length Equivalents at Selected Latitudes

Longitudinal length equivalents at selected latitudes					
Latitude	*City*	*Degree*	*Minute*	*Second*	*±0.0001°*
60°	Saint Petersburg	55.80 km	0.930 km	15.50 m	5.58 m
51° 28' 38" N	Greenwich	69.47 km	1.158 km	19.30 m	6.95 m
45°	Bordeaux	78.85 km	1.31 km	21.90 m	7.89 m
30°	New Orleans	96.49 km	1.61 km	26.80 m	9.65 m
0°	Quito	111.3 km	1.855 km	30.92 m	11.13 m

Geodetic Datum

A Geodetic system or geodetic datum is a coordinate system, and a set of reference points, used to locate places on the Earth (or similar objects). The modern definition of sea level is actually defined precisely by the datum WGS 84 from 1984 onward. Other datums are defined for other areas or at other times; ED50 was defined in 1950 over Europe and differs from WGS 84 by a few hundred metres depending on where in Europe you look. Mars has no oceans and so no sea level, but at least two martian datums have been used to locate places there.

Datums are used in geodesy, navigation, and surveying by cartographers and satellite navigation systems to translate positions indicated on maps (paper or digital) to their real position on Earth. Each starts with an ellipsoid (stretched sphere), and then defines latitude, longitude and altitude coordinates. One or more locations on the earth's surface is chosen as an anchor. For instance, at Meades Ranch, Kansas USA, 12 miles (19 km) north of Lucas, in the middle of a field, there is a bronze surveyor's disk embedded in concrete that has a dot in the middle, which is defined as 39.2240867167°N 98.5421516778°W, in the NAD83 datum (very close to WGS 84).

The difference in co-ordinates between datums is commonly referred to as *datum shift*. The datum shift between two particular datums can vary from one place to another within one country or region, and can be anything from zero to hundreds of metres (or several kilometres for some remote islands). The North Pole, South Pole and Equator may be assumed to be in different positions on different datums, so True North may be very slightly different. Different datums use different estimates for the precise shape and size of the Earth (reference ellipsoids).

Because the Earth is an imperfect ellipsoid, localised datums can give a more accurate representation of the area of coverage than WGS 84. OSGB36, for example, is a better approximation to the geoid covering the British Isles than the global WGS 84 ellipsoid. However, as the benefits of a global system outweigh the greater accuracy, the global WGS 84 datum is becoming increasingly adopted.

Horizontal datums are used for describing a point on the Earth's surface, in latitude and longitude or another coordinate system. Vertical datums measure elevations or depths.

In engineering and drafting, a *datum* is a reference point, surface, or axis on an object against which measurements are made.

Horizontal Datum

The horizontal datum is the model used to measure positions on the earth. A specific point on the earth can have substantially different coordinates, depending on the datum used to make the measurement. There are hundreds of locally-developed horizontal datums around the world, usually referenced to some convenient local reference point. Contemporary datums, based on increasingly accurate measurements of the shape of the earth, are intended to cover larger areas. The WGS 84 datum, which is almost identical to the NAD83 datum used in North America and the ETRS89 datum used in Europe, is a common standard datum.

For example, in Sydney there is a 200 metres (700 feet) difference between GPS coordinates configured in GDA (based on global standard WGS84) and AGD (used for most local maps), which is an unacceptably large error for some applications, such as surveying or site location for scuba diving.

Vertical Datum

A vertical datum is used for measuring the elevations of points on the sea level. Vertical datums are either: tidal, based on sea levels; gravimetric, based on a geoid; or geodetic, based on the same ellipsoid models of the earth used for computing horizontal datums.

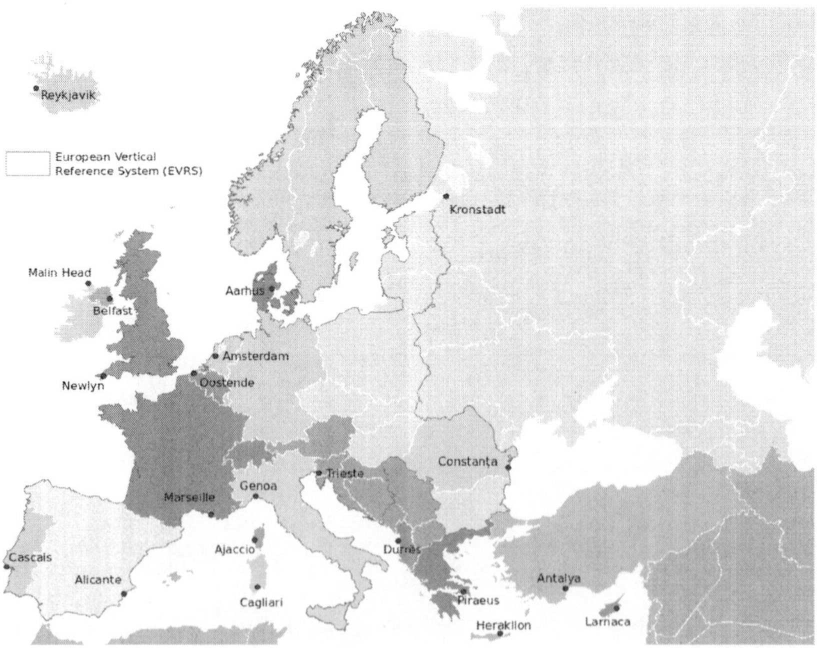

***Figure:** Vertical datums in Europe*

In common usage, elevations are often cited in height above sea level, although what "sea level" actually means is a more complex issue than might at first be thought: the height of the sea surface at any one place and time is a result of numerous effects, including waves, wind and currents, atmospheric pressure, tides, topography, and even differences in the strength of gravity due to the presence of mountains etc.

For the purpose of measuring the height of objects on land, the usual datum used is mean sea level (MSL). This is a tidal datum which is described as the arithmetic mean of the hourly water elevation taken over a specific 19 years cycle. This definition averages out tidal highs and lows (caused by the gravitational effects of the sun and the moon) and short term variations. It will not remove the effects of local gravity strength, and so the height of MSL, relative to a geodetic datum, will vary around the world, and even around one country. Countries tend to choose the mean sea level at one specific point to be used as the standard "sea level" for all mapping and surveying in that country. (For example, in Great Britain, the national vertical datum, Ordnance Datum Newlyn, is based on what was mean sea level at Newlyn in Cornwall between 1915 and 1921). However, zero elevation as defined by one country is not the same as zero elevation defined by another (because MSL is not the same everywhere), which is why locally defined vertical datums differ from one another.

A different principle is used when choosing a datum for nautical charts. For safety reasons, a mariner must be able to know the minimum depth of water that could occur at any point. For this reason, depths and tides on a nautical chart are measured relative to chart datum, which is defined to be a level below which tide rarely falls. Exactly how this is chosen depends on the tidal regime in the area being charted and on the policy of the hydrographic office producing the chart in question; a typical definition is Lowest Astronomical Tide (the lowest tide predictable from the effects of gravity), or Mean Lower Low Water (the average lowest tide of each day), although MSL is sometimes used in waters with very low tidal ranges.

Conversely, if a ship is to safely pass under a low bridge or overhead power cable, the mariner must know the minimum clearance between the masthead and the obstruction, which will occur at high tide. Consequently, bridge clearances etc. are given relative to a datum based on high tide, such as Highest Astronomical Tide or Mean High Water Springs.

Sea level does not remain constant throughout geological time, and so tidal datums are less useful when studying very long-term processes. In some situations sea level does not apply at all — for instance for mapping Mars' surface — forcing the use of a different "zero elevation", such as mean radius.

A geodetic vertical datum takes some specific zero point, and computes elevations based on the geodetic model being used, without further reference to sea levels. Usually, the starting reference point is a tide gauge, so at that point the geodetic and tidal datums might match, but due to sea level variations, the two scales may not match elsewhere. An example of a gravity-based geodetic datum is NAVD88, used in North America, which is referenced to a point in Quebec, Canada. Ellipsoid-based datums such as WGS84, GRS80 or NAD83 use a theoretical surface that may differ significantly from the geoid.

Geodetic Coordinates

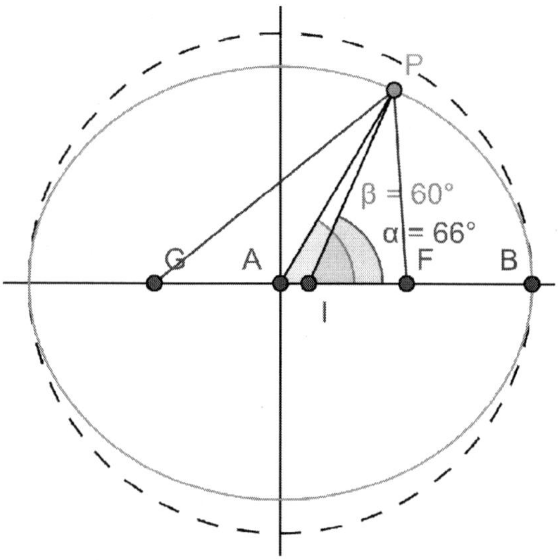

Figure: *The same position on a spheroid has a different angle for latitude depending on whether the angle is measured from the normal line segment* IP *of the ellipsoid (angle α) or the line segment* AP *from the centre (angle* β*). Note that the "flatness" of the spheroid (orange) in the image is greater than that of the Earth; as a result, the corresponding difference between the "geodetic" and "geocentric" latitudes is also exaggerated.*

In geodetic coordinates the Earth's surface is approximated by an ellipsoid and locations near the surface are described in terms of latitude (ϕ), longitude (λ) and height (*h*).

Geodetic versus Geocentric Latitude

It is important to note that geodetic latitude (ϕ) (resp. altitude) is different from geocentric latitude (ϕ') (resp. altitude). Geodetic latitude is determined by the angle between the normal of the ellipsoid and the plane of the equator, whereas geocentric latitude is determined around the centre. Unless otherwise specified latitude is geodetic latitude.

Reference Datums

A reference datum is a known and constant surface which is used to describe the location of unknown points on the earth. Since reference datums can have different radii and different centre points, a specific point on the earth can have substantially different coordinates depending on the datum used to make the measurement. There are hundreds of locally-developed reference datums around the world, usually referenced to some convenient local reference point. Contemporary datums, based on increasingly accurate measurements of the shape of the earth, are intended to cover larger areas. The most common reference Datums in use in North America are NAD27, NAD83, and WGS84.

The North American Datum of 1927 (NAD 27) is "the horizontal control datum for the United States that was defined by a location and azimuth on the Clarke spheroid of 1866, with origin at (the survey station) Meades Ranch (Kansas)." ... The geoidal height at Meades Ranch was assumed to be zero. "Geodetic positions on the North American Datum of 1927 were derived from the (coordinates of and an azimuth at Meades Ranch) through a readjustment of the triangulation of the entire network in which Laplace azimuths were introduced, and the Bowie method was used." NAD27 is a local referencing system covering North America.

The North American Datum of 1983 (NAD 83) is "The horizontal control datum for the United States, Canada, Mexico, and Central America, based on a geocentric origin and the Geodetic Reference System 1980 (GRS80). "This datum, designated as NAD 83 ...is based on the adjustment of 250,000 points including 600 satellite Doppler stations which constrain the system to a geocentric origin." NAD83 may be considered a local referencing system.

WGS 84 is the World Geodetic System of 1984. It is the reference frame used by the U.S. Department of Defence (DoD) and is defined by the National Geospatial-Intelligence Agency (NGA) (formerly the Defence Mapping Agency, then the National Imagery and Mapping Agency). WGS 84 is used by DoD for all its mapping, charting, surveying, and navigation needs, including its GPS "broadcast" and "precise" orbits. WGS 84 was defined in January 1987 using Doppler satellite surveying techniques. It was used as the reference frame for broadcast GPS Ephemerides (orbits) beginning January 23, 1987. At 0000 GMT January 2, 1994, WGS 84 was upgraded in accuracy using GPS measurements.

The formal name then became WGS 84 (G730), since the upgrade date coincided with the start of GPS Week 730. It became the reference frame for broadcast orbits on June 28, 1994. At 0000 GMT September 30, 1996 (the start of GPS Week 873), WGS 84 was redefined again and was more closely aligned with International Earth Rotation Service (IERS) frame ITRF 94. It is now formally called WGS 84 (G873). WGS 84 (G873) was adopted as the reference frame for broadcast orbits on January 29, 1997.

The WGS84 datum, which is almost identical to the NAD83 datum used in North America, is the only world referencing system in place today. WGS84 is the default standard datum for coordinates stored in recreational and commercial GPS units.

Users of GPS are cautioned that they must always check the datum of the maps they are using. To correctly enter, display, and to store map related map coordinates, the datum of the map must be entered into the GPS map datum field.

Datums Often Encountered

Latitude and longitude values can be based on different geodetic systems or datums, the most common being WGS 84, a global datum used by all GPS equipment. Other datums are significant because they were chosen by a national cartographical organisation as the best method for representing their region, and these are the datums used on printed maps. The latitude and longitude on a map may not be the same as on a GPS receiver. Coordinates from the mapping system can sometimes be roughly changed into another datum using a simple translation. For example, to convert from ETRF89 (GPS) to the Irish Grid add 49 metres to the east, and subtract 23.4 metres from the north. More generally one datum is changed into any other datum using a process called Helmert transformations. This involves converting the spherical coordinates into Cartesian coordinates and applying a seven parameter transformation (translation, three-dimensional rotation), and converting back.

In popular GIS software, data projected in latitude/longitude is often represented as a 'Geographic Coordinate System'. For example, data in latitude/longitude if the datum is the North American Datum of 1983 is denoted by 'GCS North American 1983'.

Geostationary Coordinates

Geostationary satellites (e.g., television satellites) are over the equator at a specific point on Earth, so their position related to Earth is expressed in longitude degrees only. Their latitude is always zero, that is, over the equator.

Figure of the Earth

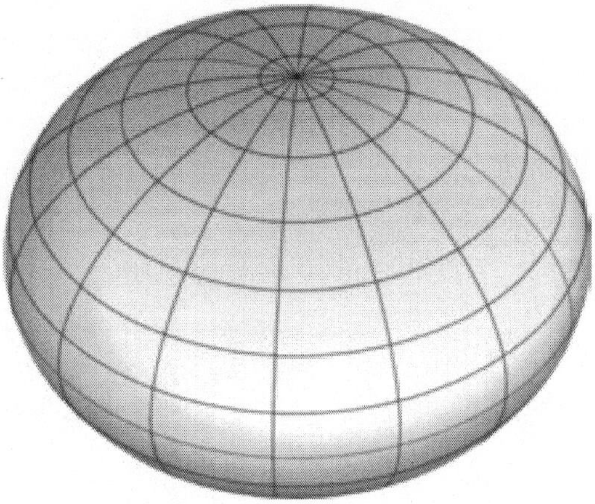

Figure: *An oblate spheroid*

The expression figure of the Earth has various meanings in geodesy according to the way it is used and the precision with which the Earth's size and shape is to be defined. While the sphere is a close approximation of the true figure of the Earth and satisfactory for many purposes, geodesists have developed a number of models to represent a closer approximation to the shape of the Earth.

Need for Models of the Figure of the Earth

The actual topographic surface is most apparent with its variety of land forms and water areas. This is, in fact, the surface on which actual Earth measurements are made. It is not suitable, however, for exact mathematical computations, because the formulas which would be required to take the irregularities into account would necessitate a prohibitive amount of computations. The topographic surface is generally the concern of topographers and hydrographers.

The Pythagorean concept of a spherical Earth offers a simple surface which is mathematically easy to deal with. Many astronomical and navigational computations use it as a surface representing the Earth. While the sphere is a close approximation of the true figure of the Earth and satisfactory for many purposes, to the geodesists interested in the measurement of long distances on the scale of continents and oceans, a more exact figure is necessary. Closer approximations range from modelling the shape of the surface of the entire Earth as an oblate spheroid or an oblate ellipsoid, to the use of spherical harmonics or local approximations in terms of local reference ellipsoids.

The idea of a planar or flat surface for Earth, however, is still sufficient for surveys of small areas, as the local topography is far more significant than the curvature. Plane-table surveys are made for relatively small areas, and no account is taken of the curvature of the Earth. A survey of a city would likely be computed as though the Earth were a plane surface the size of the city. For such small areas, exact positions can be determined relative to each other without considering the size and shape of the entire Earth.

Figure: *The curvature of Earth as seen in Valencia, Spain (Playa de la Malvarrosa)*

In the mid- to late 20th century, research across the geosciences contributed to drastic improvements in the accuracy of the figure of the Earth. The primary utility (and the motivation for funding, mainly from the military) of this improved accuracy was to provide geographical and gravitational data for the inertial guidance systems of ballistic missiles. This funding also drove the expansion of geoscientific disciplines, fostering the creation and growth of various geoscience departments at many universities.

Models of the Figure of the Earth

The models for the figure of the Earth vary in the way they are used, in their complexity, and in the accuracy with which they represent the size and shape of the Earth.

Sphere

The simplest model for the shape of the entire Earth is a sphere. The Earth's radius is the distance from Earth's centre to its surface, about 6,371 kilometres (3,959 mi). While "radius" normally is a characteristic of perfect spheres, the Earth deviates from a perfect sphere by only a third of a percent, sufficiently close to treat it as a sphere in many contexts and justifying the term "the radius of the Earth".

The concept of a spherical Earth dates back to around the 6th century BC, but remained a matter of philosophical speculation until the 3rd century BC. The first scientific estimation of the radius of the earth was given by Eratosthenes about 240 BC, with estimates of the accuracy of Eratosthenes's measurement ranging from 2% to 15%.

The Earth is only approximately spherical, so no single value serves as its natural radius. Distances from points on the surface to the centre range from 6,353 km to 6,384 km (3,947 – 3,968 mi). Several different ways of modelling the Earth as a sphere each yield a mean radius of 6,371 kilometres (3,959 mi). Regardless of the model, any radius falls between the polar minimum of about 6,357 km and the equatorial maximum of about 6,378 km (3,950 – 3,963 mi).

Ellipsoid of Revolution

Since the Earth is flattened at the poles and bulges at the equator, geodesy represents the shape of the earth with an oblate spheroid. The oblate spheroid, or oblate ellipsoid, is an ellipsoid of revolution obtained by rotating an ellipse about its shorter axis. It is the regular geometric shape that most nearly approximates the shape of the Earth. A spheroid describing the figure of the Earth or other celestial body is called a reference ellipsoid. The reference ellipsoid for Earth is called an Earth ellipsoid.

An ellipsoid of revolution is uniquely defined by two numbers: two dimensions, or one dimension and a number representing the difference between the two dimensions. Geodesists, by convention, use the semi-major axis and flattening. The size is represented by the radius at the equator (the semimajor axis of the cross-sectional ellipse) and designated by the letter a. The shape of the ellipsoid is given by the flattening, f, which indicates how much the ellipsoid departs from spherical. (In practice, the two defining numbers are usually the equatorial radius and the reciprocal of the flattening, rather than the flattening itself; for the WGS84 spheroid used by today's GPS systems, the reciprocal of the flattening is set at 298.257223563 exactly.)

The difference between a sphere and a reference ellipsoid for Earth is small, only about one part in 300. Historically flattening was computed from grade measurements. Nowadays geodetic networks and satellite geodesy are used. In practice, many reference ellipsoids have been developed over the centuries from different surveys. The flattening value varies slightly from one reference ellipsoid to another, reflecting local conditions and whether the reference ellipsoid is intended to model the entire Earth or only some portion of it.

A sphere has a single radius of curvature, which is simply the radius of the sphere. More complex surfaces have radii of curvature that vary over the surface. The radius of curvature describes the radius of the sphere that best approximates the surface at that point. Oblate ellipsoids have constant radius of curvature east to west along parallels, if a graticule is drawn on the surface, but varying curvature in any other direction. For an oblate ellipsoid, the polar radius of curvature r_p is larger than the equatorial

$$r_p = \frac{a^2}{b},$$

because the pole is flattened: the flatter the surface, the larger the sphere must be to approximate it. Conversely, the ellipsoid's north-south radius of curvature at the equator r_e is smaller than the polar

$$r_e = \frac{b^2}{a}$$

where a is the distance from the centre of the ellipsoid to the equator (semi-major axis), and b is the distance from the centre to the pole. (semi-minor axis)

More Complicated Shapes

The possibility that the Earth's equator is an ellipse rather than a circle and therefore that the ellipsoid is triaxial has been a matter of scientific controversy for many years. Modern technological developments have furnished new and rapid methods for data collection and since the launch of *Sputnik 1*, orbital data have been used to investigate the theory of ellipticity.

A second theory, more complicated than triaxiality, proposed that observed long periodic orbital variations of the first Earth satellites indicate an additional depression at the south pole accompanied by a bulge of the same degree at the north pole. It is also contended that the northern middle latitudes were slightly flattened and the southern middle latitudes bulged in a similar amount. This concept suggested a slightly pear-shaped Earth and was the subject of much public discussion. Modern geodesy tends to retain the ellipsoid of revolution and treat triaxiality and pear shape as a part of the geoid Figure: they are represented by the spherical harmonic coefficients C_{22}, S_{22} and C_{30}, respectively, corresponding to degree and order numbers 2.2 for the triaxiality and 3.0 for the pear shape.

Geoid

It was stated earlier that measurements are made on the apparent or topographic surface of the Earth and it has just been explained that computations are performed on an ellipsoid. One other surface is involved in geodetic measurement: the geoid. In geodetic surveying, the computation of the geodetic coordinates of points is commonly performed on a reference ellipsoid closely approximating the size and shape of the Earth in the area of the survey. The actual measurements made on the surface of the Earth with certain instruments are however referred to the geoid. The ellipsoid is a mathematically defined regular surface with specific dimensions. The geoid, on the other hand, coincides with that surface to which the oceans would conform over the entire Earth if free to adjust to the combined effect of the Earth's mass attraction (gravitation) and the centrifugal force of the Earth's rotation. As a result of the uneven distribution of the Earth's mass, the geoidal surface is irregular and, since the ellipsoid is a regular surface, the separations between the two, referred to as geoid undulations, geoid heights, or geoid separations, will be irregular as well.

The geoid is a surface along which the gravity potential is everywhere equal and to which the direction of gravity is always perpendicular. The latter is particularly important because optical instruments containing gravity-reference leveling devices are commonly used to make geodetic measurements. When properly adjusted, the vertical axis of the instrument coincides with the direction of gravity and is, therefore, perpendicular to the geoid. The angle between the plumb line which is perpendicular to the geoid (sometimes called "the vertical") and the perpendicular to the ellipsoid (sometimes called "the ellipsoidal normal") is defined as the deflection of the vertical. It has two components: an east-west and a north-south component.

Earth Rotation and Earth's Interior

Determining the exact figure of the Earth is not only a geodetic operation or a task of geometry, but is also related to geophysics. Without any idea of the Earth's interior, we can state a "constant density" of 5.515 g/cm^3 and, according to theoretical arguments, such a body rotating like the Earth would have a flattening of 1:230.

In fact the measured flattening is 1:298.25, which is more similar to a sphere and a strong argument that the Earth's core is *very compact*.

Therefore the density must be a function of the depth, reaching from about 2.7 g/cm^3 at the surface (rock density of granite, limestone etc.) up to approximately 15 within the inner core. Modern seismology yields a value of 16 g/cm^3 at the centre of the Earth.

Global and Regional Gravity Field

Also with implications for the physical exploration of the Earth's interior is the gravitational field, which can be measured very accurately at the surface and remotely by satellites. True vertical generally does not correspond to theoretical vertical (deflection ranges from 2" to 50") because topography and all *geological masses* disturb the gravitational field. Therefore the gross structure of the earth's crust and mantle can be determined by geodetic-geophysical models of the subsurface.

Spherical Coordinate System

A spherical coordinate system is a coordinate system for three-dimensional space where the position of a point is specified by three numbers: the radial distance of that point from a fixed origin, its polar angle measured from a fixed zenith direction, and the azimuth angle of its orthogonal projection on a reference plane that passes through the origin and is orthogonal to the zenith, measured from a fixed reference direction on that plane.

The radial distance is also called the radius or radial coordinate. The polar angle may be called co-latitude, zenith angle, normal angle, or inclination angle.

The use of symbols and the order of the coordinates differs between sources. In one system frequently encountered in physics (r, θ, φ) gives the radial distance, polar angle, and azimuthal angle, whereas in another system used in many mathematics books (r, θ, φ) gives the radial distance, azimuthal angle, and polar angle. In both systems ρ is often used instead of r. Other conventions are also used, so great care needs to be taken to check which one is being used.

A number of different spherical coordinate systems following other conventions are used outside mathematics. In a geographical coordinate system positions are measured in latitude, longitude and height or altitude. There are a number of different celestial coordinate systems based on different fundamental planes and with different terms for the various coordinates. The spherical coordinate systems used in mathematics normally use radians rather than degrees and measure the azimuthal angle counter-clockwise rather than clockwise. The inclination angle is often replaced by the elevation angle measured from the reference plane. Elevation angle of zero is at the horizon.

The concept of spherical coordinates can be extended to higher-dimensional spaces and are then referred to as hyperspherical coordinates.

Definition

To define a spherical coordinate system, one must choose two orthogonal directions, the *zenith* and the *azimuth reference*, and an *origin* point in space. These choices determine a reference plane that contains the origin and is perpendicular to the zenith. The spherical coordinates of a point *P* are then defined as follows:

- The *radius* or *radial distance* is the Euclidean distance from the origin *O* to *P*.
- The *inclination* (or *polar angle*) is the angle between the zenith direction and the line segment *OP*.
- The *azimuth* (or *azimuthal angle*) is the signed angle measured from the azimuth reference direction to the orthogonal projection of the line segment *OP* on the reference plane.

The sign of the azimuth is determined by choosing what is a *positive* sense of turning about the zenith. This choice is arbitrary, and is part of the coordinate system's definition.

The *elevation* angle is 90 degrees ($\pi/2$ radians) minus the inclination angle.

If the inclination is zero or 180 degrees (π radians), the azimuth is arbitrary. If the radius is zero, both azimuth and inclination are arbitrary.

In linear algebra, the vector from the origin *O* to the point *P* is often called the *position vector* of *P*.

Conventions

Several different conventions exist for representing the three coordinates, and for the order in which they should be written. The use of (r, θ, φ) to denote radial distance, inclination (or elevation), and azimuth, respectively, is common practice in physics, and is specified by ISO standard 80000-2 :2009, and earlier in ISO 31-11.

However, some authors (including mathematicians) use φ for inclination (or elevation) and θ for azimuth, which "provides a logical extension of the usual polar coordinates notation". Some authors may also list the azimuth before the inclination (or elevation), and/or use ρ (rho) instead of r for radial distance. Some combinations of these choices result in a left-handed coordinate system. The standard convention (r, θ, φ) conflicts with the usual notation for the two-dimensional polar coordinates, where θ is often used for the azimuth. It may also conflict with the notation used for three-dimensional cylindrical coordinates.

The angles are typically measured in degrees (°) or radians (rad), where 360° = 2π rad. Degrees are most common in geography, astronomy, and engineering, whereas radians are commonly used in mathematics and theoretical physics. The unit for radial distance is usually determined by the context.

When the system is used for physical three-space, it is customary to use positive sign for azimuth angles that are measured in the counter-clockwise sense from the reference direction on the reference plane, as seen from the zenith side of the plane. This convention is used, in particular, for geographical coordinates, where the "zenith" direction is north and positive azimuth (longitude) angles are measured eastwards from some prime meridian.

Major conventions		
coordinates	*corresponding local geographical directions* (Z, X, Y)	*right/left-handed*
$(r, \theta_{\text{inc}}, \phi_{\text{az,right}})$	(U, S, E)	right
$(r, \phi_{\text{az,right}}, \theta_{\text{el}})$	(U, E, N)	right
$(r, \theta_{\text{el}}, \phi_{\text{az,right}})$	(U, N, E)	left
Note: easting (E), northing (N), upwardness (U). Local azimuth angle would be measured, e.g., counterclockwise from Sto E in the case of (U, S, E).		

Unique Coordinates

Any spherical coordinate triplet (r, θ, φ) specifies a single point of three-dimensional space. On the other hand, every point has infinitely many equivalent spherical coordinates. One can add or subtract any number of full turns to either angular measure without changing the angles themselves, and therefore without changing the point. It is also convenient, in many contexts, to allow negative radial distances, with the convention that ($-r$, θ, φ) is equivalent to (r, θ + 180°, φ) for any r, θ, and φ. Moreover, (r, −θ, φ) is equivalent to (r, θ, φ + 180°).

If it is necessary to define a unique set of spherical coordinates for each point, one may restrict their ranges. A common choice is:

$$r \geq 0$$

$$0° \leq \theta \leq 180° \ (\pi \text{ rad})$$

$$0° \leq \varphi < 360° \ (2\pi \text{ rad})$$

However, the azimuth φ is often restricted to the interval (–180°, +180°], or ($-\pi$, $+\pi$] in radians, instead of [0, 360°). This is the standard convention for geographic longitude.

The range [0°, 180°] for inclination is equivalent to [–90°, +90°] for elevation (latitude). Even with these restrictions, if θ is zero or 180° (elevation is 90° or –90°) then the azimuth angle is arbitrary; and if r is zero, both azimuth and inclination/elevation are arbitrary. To make the coordinates unique, one can use the convention that in these cases the arbitrary coordinates are zero.

Plotting

To plot a point from its spherical coordinates (r, θ, φ), where θ is inclination, move r units from the origin in the zenith direction, rotate by θ about the origin towards the azimuth reference direction, and rotate by φ about the zenith in the proper direction.

Applications

The geographic coordinate system uses the azimuth and elevation of the spherical coordinate system to express locations on Earth, calling them respectively longitude and latitude. Just as the two-dimensional Cartesian coordinate system is useful on the plane, a two-dimensional spherical coordinate system is useful on the surface of a sphere. In this system, the sphere is taken as a unit sphere, so the radius is unity and can generally be ignored. This simplification can also be very useful when dealing with objects such as rotational matrices.

Spherical coordinates are useful in analyzing systems that have some degree of symmetry about a point, such as volume integrals inside a sphere, the potential energy field surrounding a concentrated mass or charge, or global weather simulation in a planet's atmosphere. A sphere that has the Cartesian equation $x^2 + y^2 + z^2 = c^2$ has the simple equation $r = c$ in spherical coordinates.

Two important partial differential equations that arise in many physical problems, Laplace's equation and the Helmholtz equation, allow a separation of variables in spherical coordinates. The angular portions of the solutions to such equations take the form of spherical harmonics.

Another application is ergonomic design, where r is the arm length of a stationary person and the angles describe the direction of the arm as it reaches out.

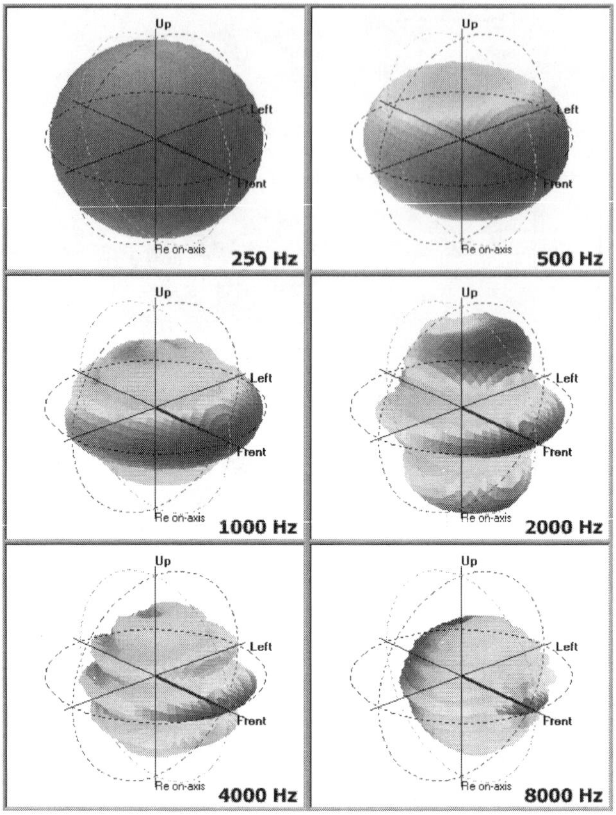

Figure: *The output pattern of an industrial loudspeaker shown using spherical polar plots taken at six frequencies*

Three dimensional modelling of loudspeaker output patterns can be used to predict their performance. A number of polar plots are required, taken at a wide selection of frequencies, as the pattern changes greatly with frequency. Polar plots help to show that many loudspeakers tend toward omnidirectionality at lower frequencies.

The spherical coordinate system is also commonly used in 3D game development to rotate the camera around the player's position.

Coordinate System Conversions

As the spherical coordinate system is only one of many three-dimensional coordinate systems, there exist equations for converting coordinates between the spherical coordinate system and others.

Cartesian Coordinates

The spherical coordinates of a point in the ISO convention (*radius* r, *inclination* θ, *azimuth* φ) can be obtained from its Cartesian coordinates (x, y, z) by the formulae

$$r = \sqrt{x^2 + y^2 + z^2}$$

$$\theta = \arccos\left(\frac{z}{\sqrt{x^2 + y^2 + z^2}}\right)$$

$$\varphi = \arctan\left(\frac{y}{x}\right)$$

The inverse tangent denoted in $\varphi = \arctan(y/x)$ must be suitably defined, taking into account the correct quadrant of (x,y).

Alternatively, the conversion can be considered as two sequential rectangular to polar conversions: the first in the Cartesian x–y plane from (x,y) to (R,φ), where R is the projection of r onto the x–y plane, and the second in the Cartesian z–R plane from (z,R) to (r,θ). The correct quadrants for φ and θ are implied by the correctness of the planar rectangular to polar conversions.

These formulae assume that the two systems have the same origin, that the spherical reference plane is the Cartesian x–y plane, that θ is inclination from the z direction, and that the azimuth angles are measured from the Cartesian x axis (so that the y axis has $\varphi = +90°$). If θ measures elevation from the reference plane instead of inclination from the zenith the across above becomes an arcsin, and the cos θ and sin θ below become switched.

Conversely, the Cartesian coordinates may be retrieved from the spherical coordinates (*radius* r, *inclination* θ, *azimuth* φ), where $r \in [0, \infty), \theta \in [0, \pi], \varphi \in [0, 2\pi)$, by:

$$x = r\sin\theta\cos\varphi$$

$$y = r\sin\theta\sin\varphi$$

$$z = r\cos\theta$$

Geographic Coordinates

To a first approximation, the geographic coordinate system uses elevation angle (latitude) in degrees north of the equator plane, in the range $-90° \le \varphi \le 90°$, instead of inclination. Latitude is either geocentric latitude, measured at the Earth's centre and designated variously by ø, q, $\varphi 2$, φ_c, φ_g or geodetic latitude, measured by the observer's local vertical, and commonly designated φ. The azimuth angle (longitude), commonly denoted by λ, is measured in degrees east or west from some conventional reference meridian (most commonly the IERS Reference

Meridian), so its domain is $-180° \leq \lambda \leq 180°$. For positions on the Earth or other solid celestial body, the reference plane is usually taken to be the plane perpendicular to the axis of rotation.

The polar angle, which is 90° minus the latitude and ranges from 0 to 180°, is called colatitude in geography.

Instead of the radial distance, geographers commonly use altitude above some reference surface, which may be the sea level or "mean" surface level for planets without liquid oceans. The radial distance r can be computed from the altitude by adding the mean radius of the planet's reference surface, which is approximately 6,360 ± 11 km for Earth.

However, modern geographical coordinate systems are quite complex, and the positions implied by these simple formulae may be wrong by several kilometres. The precise standard meanings of latitude, longitude and altitude are currently defined by the World Geodetic System (WGS), and take into account the flattening of the Earth at the poles (about 21 km) and many other details.

In astronomy there are a series of spherical coordinate systems that measure the elevation angle from different fundamental planes. These reference planes are the observer's horizon, the celestial equator (defined by the Earth's rotation), the plane of the ecliptic (defined by Earth's orbit around the sun), and the galactic equator (defined by the rotation of the galaxy).

World Geodetic System

The World Geodetic System (WGS) is a standard for use in cartography, geodesy, and navigation. It comprises a standard coordinate system for the Earth, a standard spheroidal reference surface (the *datum* or *reference ellipsoid*) for raw altitude data, and a gravitational equipotential surface (the *geoid*) that defines the *nominal sea level*.

The latest revision is WGS 84 (aka WGS 1984, EPSG:4326), established in 1984 and last revised in 2004. Earlier schemes included WGS 72, WGS 66, and WGS 60. WGS 84 is the reference coordinate system used by the Global Positioning System.

Main Parameters

The coordinate origin of WGS 84 is meant to be located at the Earth's centre of mass; the error is believed to be less than 2 cm.

The WGS 84 meridian of zero longitude is the IERS Reference Meridian, 5.31 arc seconds or 102.5 metres (336.3 ft) east of the Greenwich meridian at the latitude of the Royal Observatory.

The WGS 84 datum surface is an oblate spheroid (ellipsoid) with major (equatorial) radius $a = 6378137$ m at the equator and flattening $f = 1/298.257223563$. The polar semi-minor axis b then equals a times $(1–f)$, or 6356752.3142 m.

Currently, WGS 84 uses the EGM96 (Earth Gravitational Model 1996) geoid, revised in 2004. This geoid defines the nominal sea level surface by means of a spherical harmonics series of degree 360 (which provides about 100 km horizontal resolution). The deviations of the EGM96 geoid from the WGS 84 reference ellipsoid range from about "105 m to about +85 m. EGM96 differs from the original WGS 84 geoid, referred to as EGM84.

History

Efforts to supplement the various national surveying systems began in the 19th century with F.R. Helmert's famous books *Mathematische und Physikalische Theorien der Physikalischen Geodäsie* (*Mathematical and Physical Theory of Physical Geodesy*). Austria and Germany founded the *Zentralbüro für die Internationale Erdmessung* (Central Bureau of International Geodesy), and a series of global ellipsoids of the Earth were derived.

A unified geodetic system for the whole world became essential in the 1950s for several reasons:

- International space science and the beginning of astronautics.
- The lack of inter-continental geodetic information.
- The inability of the large geodetic systems, such as European Datum (ED50), North American Datum (NAD), and Tokyo Datum (TD), to provide a worldwide geo-data basis
- Need for global maps for navigation, aviation, and geography.
- Western Cold War preparedness necessitated a standardised, NATO-wide geospatial reference system, in accordance with the NATO Standardisation Agreement

In the late 1950s, the United States Department of Defence, together with scientists of other institutions and countries, began to develop the needed world system to which geodetic data could be referred and compatibility established between the coordinates of widely separated sites of interest. Efforts of the U.S. Army, Navy and Air Force were combined leading to the DoD World Geodetic System 1960 (WGS 60). The term *datum* as used here refers to a smooth surface somewhat arbitrarily defined as zero elevation, consistent with a set of surveyor's measures of distances between various stations, and differences in elevation, all reduced to a grid of latitudes, longitudes, and elevations. Heritage surveying methods found elevation differences from a local horizontal determined by the spirit level, plumb line, or an equivalent device that depends on the local gravity field.

As a result, the elevations in the data are referenced to the geoid, a surface that is not readily found using satellite geodesy. The latter observational method is more suitable for global mapping. Therefore, a motivation, and a substantial problem in the WGS and similar work is to patch together data that were not only made separately, for different regions, but to re-reference the elevations to an ellipsoid model rather than to the geoid.

In accomplishing WGS 60, a combination of available surface gravity data, astro-geodetic data and results from HIRAN and Canadian SHORAN surveys were used to define a best-fitting ellipsoid and an earth-centred orientation for each of initially selected datum. (Every datum is relatively oriented with respect to different portions of the geoid by the astro-geodetic methods already described.) The sole contribution of satellite data to the development of WGS 60 was a value for the ellipsoid flattening which was obtained from the nodal motion of a satellite.

Prior to WGS 60, the U.S. Army and U.S. Air Force had each developed a world system by using different approaches to the gravimetric datum orientation method. To determine their gravimetric orientation parameters, the Air Force used the mean of the differences between the gravimetric and astro-geodetic deflections and geoid heights (undulations) at specifically selected stations in the areas of the major datums. The Army performed an adjustment to minimize the difference between astro-geodetic and gravimetric geoids. By matching the relative astro-geodetic geoids of the selected datums with an earth-centred gravimetric geoid, the selected datums were reduced to an earth-centred orientation. Since the Army and Air Force systems agreed remarkably well for the NAD, ED and TD areas, they were consolidated and became WGS 60.

The United States Department of Defence World Geodetic System 1966

Steps to the improvement of a global system were the Astrogeoid of Irene Fischer and the astronautic Mercury datum. In January 1966, a World Geodetic System Committee composed of representatives from the United States Army, Navy and Air Force, was charged with the responsibility of developing an improved WGS needed to satisfy mapping, charting and geodetic requirements. Additional surface gravity observations, results from the extension of triangulation and trilateration networks, and large amounts of Doppler and optical satellite data had become available since the development of WGS 60.

Using the additional data and improved techniques, WGS 66 was produced which served DoD needs for about five years after its implementation in 1967. The defining parameters of the WGS 66 Ellipsoid were the flattening (1/298.25), determined from satellite data and the semimajor axis (6,378,145 metres), determined from a combination of Doppler satellite and astro-geodetic data. A worldwide 5° × 5° mean free air gravity anomaly field provided the basic data for producing the WGS 66 gravimetric geoid. Also, a geoid referenced to the WGS 66 Ellipsoid was derived from available astrogeodetic data to provide a detailed representation of limited land areas.

The United States Department of Defence World Geodetic System 1972

After an extensive effort extending over a period of approximately three years, the Department of Defence World Geodetic System 1972 was completed. Selected satellite, surface gravity and astrogeodetic data available through 1972 from both DoD and non-DoD sources were used in a Unified WGS Solution (a large scale least squares adjustment). The results of the adjustment consisted of corrections to initial station coordinates and coefficients of the gravitational field.

The largest collection of data ever used for WGS purposes was assembled, processed and applied in the development of WGS 72. Both optical and electronic satellite data were used. The electronic satellite data consisted, in part, of Doppler data provided by the U.S. Navy and cooperating non-DoD satellite tracking stations established in support of the Navy's Navigational Satellite System (NNSS).

Doppler data was also available from the numerous sites established by GEOCEIVERS during 1971 and 1972. Doppler data was the primary data source for WGS 72. Additional electronic satellite data was provided by the SECOR (Sequential Collation of Range) Equatorial Network completed by the U.S. Army in 1970. Optical satellite data from the Worldwide Geometric Satellite Triangulation Program was provided by the BC-4 camera system. Data from the Smithsonian Astrophysical Observatory was also used which included camera (Baker Nunn) and some laser ranging.

Doppler Satellite Ground Stations Providing Data for WGS 72 Development.

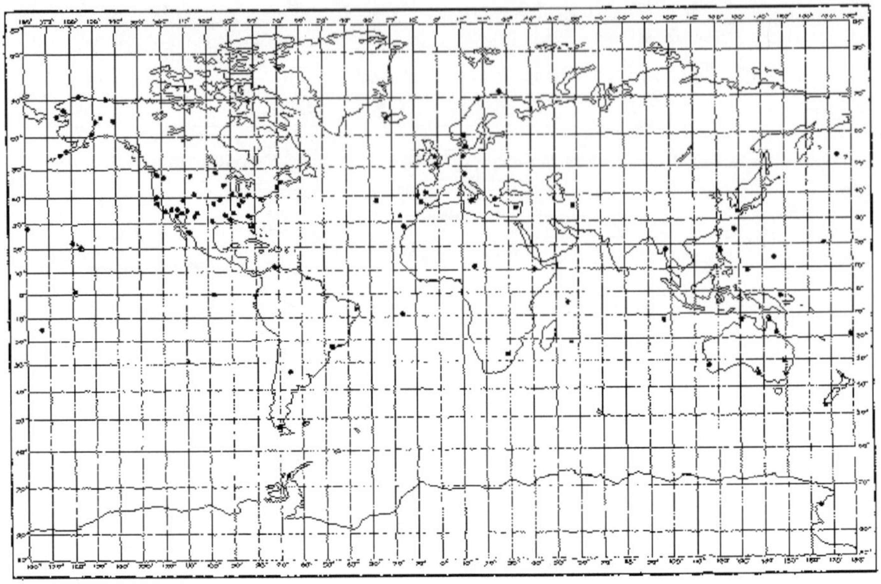

***Figure:** Doppler satellite ground stations providing data for WGS 72 development*

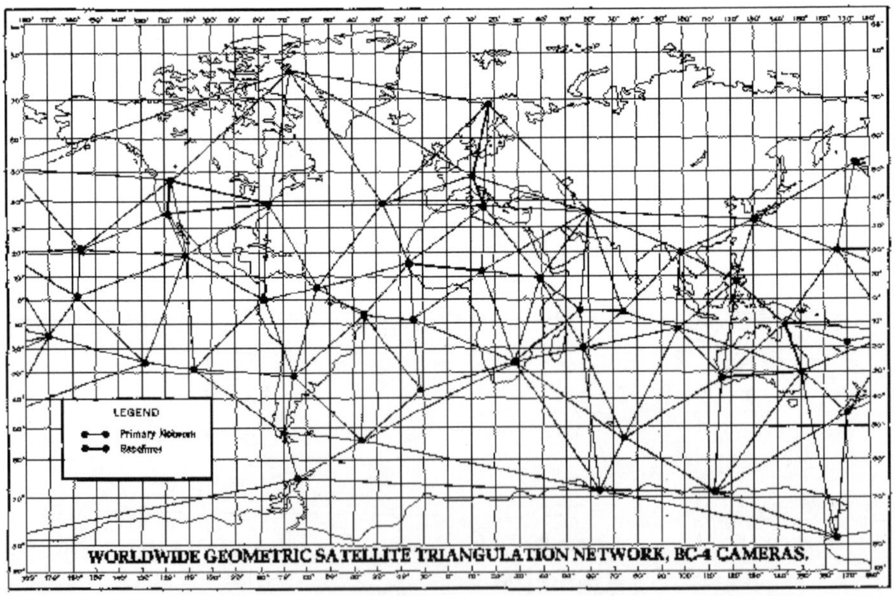

***Figure:** Worldwide geometric satellite triangulation network, BC-4 cameras*

The surface gravity field used in the Unified WGS Solution consisted of a set of 410 $10° \times 10°$ equal area mean free air gravity anomalies determined solely from terrestrial data. This gravity field includes mean anomaly values compiled directly from observed gravity data wherever the latter was available in sufficient quantity. The value for areas of sparse or no observational data were developed from geophysically compatible gravity approximations using gravity-geophysical correlation techniques. Approximately 45 percent of the 410 mean free air gravity anomaly values were determined directly from observed gravity data.

The astrogeodetic data in its basic form consists of deflection of the vertical components referred to the various national geodetic datums. These deflection values were integrated into astrogeodetic geoid charts referred to these national datums. The geoid heights contributed to the Unified WGS Solution by providing additional and more detailed data for land areas. Conventional ground survey data was included in the solution to enforce a consistent adjustment of the coordinates of neighbouring observation sites of the BC-4, SECOR, Doppler and Baker-Nunn systems. Also, eight geodimetre long line precise traverses were included for the purpose of controlling the scale of the solution.

The Unified WGS Solution, as stated above, was a solution for geodetic positions and associated parameters of the gravitational field based on an optimum combination of available data. The WGS 72 ellipsoid parameters, datum shifts and other associated constants were derived separately. For the unified solution, a normal equation matrix was formed based on each of the mentioned data sets. Then, the individual normal equation matrices were combined and the resultant matrix solved to obtain the positions and the parameters.

The value for the semimajor axis (a) of the WGS 72 Ellipsoid is 6 378 135 metres. The adoption of an a-value 10 metres smaller than that for the WGS 66 Ellipsoid was based on several calculations and indicators including a combination of satellite and surface gravity data for position and gravitational field determinations. Sets of satellite derived station coordinates and gravimetric deflection of the vertical and geoid height data were used to determine local-to-geocentric datum shifts, datum rotation parameters, a datum scale parameter and a value for the semimajor axis of the WGS Ellipsoid. Eight solutions were made with the various sets of input data, both from an investigative point of view and also because of the limited number of unknowns which could be solved for in any individual solution due to computer limitations. Selected Doppler satellite tracking and astro-geodetic datum orientation stations were included in the various solutions. Based on these results and other related studies accomplished by the Committee, an a-value of 6 378 135 metres and a flattening of 1/298.26 were adopted.

In the development of local-to WGS 72 datum shifts, results from different geodetic disciplines were investigated, analyzed and compared. Those shifts adopted were based primarily on a large number of Doppler TRANET and GEOCEIVER station coordinates which were available worldwide. These coordinates had been determined using the Doppler point positioning method.

A New World Geodetic System: WGS 84

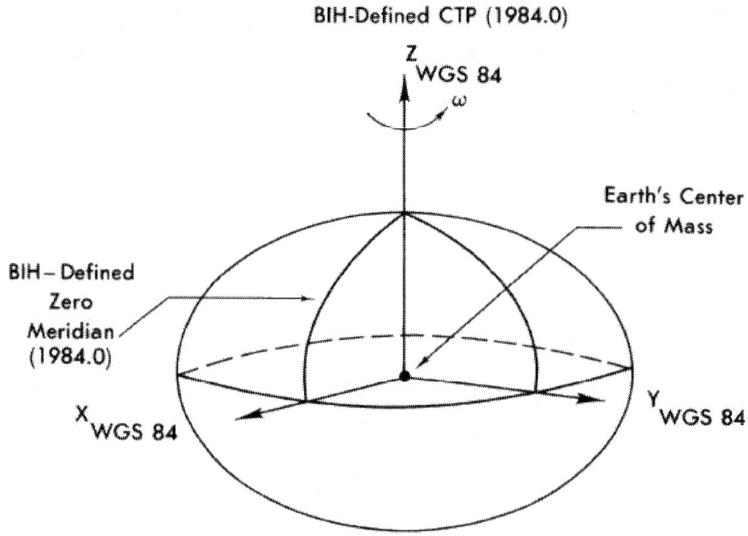

Figure: WGS 84 reference frame.

In the early 1980s the need for a new world geodetic system was generally recognised by the geodetic community, also within the US Department of Defence. WGS 72 no longer provided sufficient data, information, geographic coverage, or product accuracy for all then-current and anticipated applications.

The means for producing a new WGS were available in the form of improved data, increased data coverage, new data types and improved techniques. GRS 80 parameters together with available Doppler, satellite laser ranging and Very Long Baseline Interferometry (VLBI) observations constituted significant new information. An outstanding new source of data had become available from satellite radar altimetry. Also available was an advanced least squares method called collocation which allowed for a consistent combination solution from different types of measurements all relative to the Earth's gravity field, i.e. geoid, gravity anomalies, deflections, dynamic Doppler, etc.

The new World Geodetic System was called WGS 84. It is currently the reference system being used by the Global Positioning System. It is geocentric and globally consistent within ±1 m. Current geodetic realisations of the geocentric reference system family International Terrestrial Reference System (ITRS) maintained by the IERS are geocentric, and internally consistent, at the few-cm level, while still being metre-level consistent with WGS 84.

The WGS 84 originally used the GRS 80 reference ellipsoid, but has undergone some minor refinements in later editions since its initial publication. Most of these refinements are important for high-precision orbital calculations for satellites but have little practical effect on typical topographical uses. The following table lists the primary ellipsoid parameters.

Ellipsoid reference	*Semi-major axis a*	*Semi-minor axis b*	*Inverse flattening (1/f)*
GRS 80	6 378 137.0 m	≈6 356 752.314 140 m	298.257 222 100 882 711...
WGS 84	6 378 137.0 m	≈6 356 752.314 245 m	298.257 223 563

The very small difference in the flattening thus results in a tiny difference of 0.105 mm in the semi polar axis.

Longitudes on WGS 84

WGS 84 uses the IERS Reference Meridian as defined by the Bureau International de l'Heure, which was defined by compilation of star observations in different countries. The mean of this data caused a shift of about 100 metres east away from the Prime Meridian at Greenwich, UK.

The longitude positions on WGS 84 agree with those on the older North American Datum 1927 at roughly 85° longitude west, in the east-central United States.

Chapter 5

Point Pattern Analysis

Point pattern analysis (PPA) is the study of the spatial arrangements of points in (usually 2-dimensional) space. The simplest formulation is a set $X = \{x - D\}$ where D, which can be called the 'study region,' is a subset of R^n, a n-dimensional Euclidean space.

Description

The easiest way to visualize a 2-D point pattern is a map of the locations, which is simply a scatterplot but with the provision that the axes are equally scaled. If D is not the boundary of the map then it should also be indicated. An empirical definition of D would be the convex hull of the points, or at least their bounding box, a matrix of the ranges of the coordinates. Another straightforward way to visualize the points is a 2D histogram (sometimes called a quadrats) that bins the points into rectangular regions. A benefit of quadrat analysis is that it forces the analysis to take into account possible scales within which statistically significant inhomogeneities may be occurring.

Modelling

The null model for point patterns is complete spatial randomness (CSR), modelled as a Poisson process in R^n, which implies that the number of points in any arbitrary region A in D will be proportional to the area or volume of A. Exploring models is generally iterative: if CSR is accepted not much more can be said, but if rejected, there are two avenues. First, one must decide which models are worth exploring, such as investigations of clustering, density, trends, etc. And for each of these models there are appropriate scale ranges, from the finest, which essentially mirrors the point pattern, to the coarsest, which aggregates *D*. It is generally interesting to explore a range of scales within these limits. A particularly robust model of clustered point patterns is diffusion, which can also be thought of as the trajectory of a point doing a random walk.

Estimation

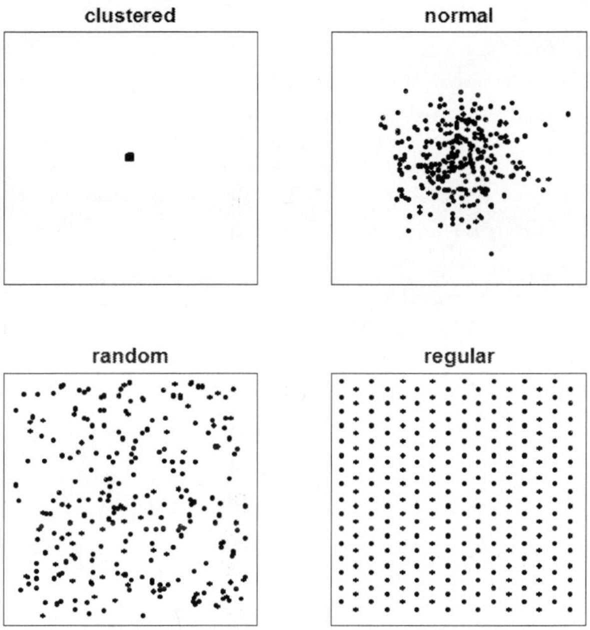

Figure: *Four patterns of 256 points*

A fundamental problem of PPA is inferring whether a given arrangement is merely random or the result of some process. The picture illustrates patterns of 256 points using four point processes. The clustered process results in all points having the same location. Popular models are those based on simple circles and ellipses, inter-point (and especially nearest neighbour) distances, quadrats, and intensity functions. Each model yields estimates (that can increase insights into the underlying real-world processes) as well as associated goodness-of-fit diagnostics.

Applications

PPA has applications in a wide range of areas, including astronomy, geography, ecology, biology, and epidemiology. A few topics in the last area are discussed here.

1. A case control study compares the point patterns of organisms both with and without some condition to determine if there were significant differences in their arrangements.

2. Environmental exposure examines the locations of cases and possible sources (e.g. of pollution or carcinogens).
3. Contagion explores the temporal unfolding of the pattern, asking about such phenomena as the location of the 'index case.'
4. Examination of infection compares the arrangements of parasites and hosts (predators and prey, agents and organisms).
5. Analysis of the regularity of retinal mosaics, particularly as a quantitative tool to understand development of the retina.

Point pattern analysis is the evaluation of the pattern, or distribution, of a set of points on a surface. It can refer to the actual spatial or temporal location of these points or also include data from point sources. It is one of the most fundamental concepts in geography and spatial analysis. Point pattern analysis is particularly appropriate for population science, as human statistics often come in point or single event form. As in many spatial analysis techniques, distribution across space can also be conceived of and analyzed as distribution across time: 'study area' can be replaced with 'study time' and distance from a point in space can be replaced with distance from a point in time. A range of methods can be applied to point pattern analysis, ranging from the simple to the complex. There are many similarities between the statistics used for point pattern analysis and those used in number analysis, although point pattern analysis also includes some speciality techniques. In this article I will discuss some of the more common methods and how they can be applied to problems in population science.

Descriptive Statistics: In its most basic form, we can think of point pattern analysis as an attempt to analyze the occurrence of points in a particular space. Often the first question asked is simply, how many points are there? For example, how many hospitals are there in a neighbourhood of a city? How does that compare to a differing neighbourhood, or that city as a whole? This is called studying the point's frequency, intensity, or abundance. To answer this and some other basic questions we can use the simple descriptive statistics that we would use for a numerical datasets: Count, Mean, Median, and Standard Deviation. Applying these, we can describe how dense a pattern is, where the centre of a set of points is, and how dispersed these points are.

Frequency and Density: Frequency is the most basic way of evaluating a spatial pattern and is simply counting the number of points in your study area. To get density, you divide this total by unit of area or time in whatever units you deem are most appropriate: Hospitals per square mile, violent crimes per month, etc. Frequency and density should almost always be determined at the start of your analysis.

Descriptive Spatial Statistics: Many of the standard descriptive statistics can be applied or slightly modified to describe spatial data. Mean centre is simply the mean of the X and the mean of the Y coordinates for your set of points, giving you the middle of your point pattern. Median Centre is a slightly different way to calculate a middle (actually quite complicated to compute) and is the point in a pattern which minimizes the distance between itself and all other points.

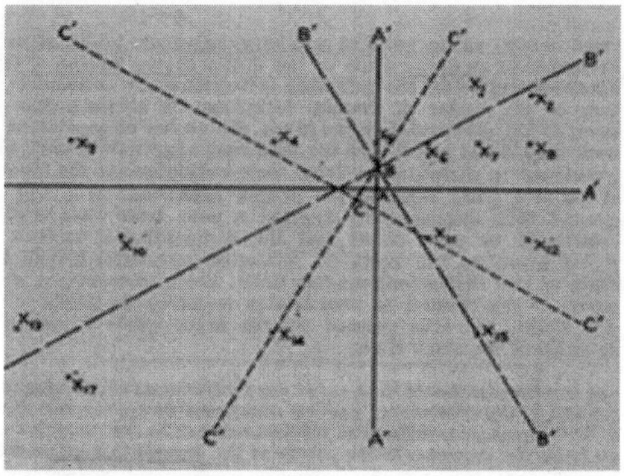

Standard Distance Deviation: is the standard deviation of the distance of each point from the mean centre. It is the spatial equivalent of standard deviation and likewise provides one with an idea of the deviation or variance of a dataset as a whole, or the deviation of a particular point from the rest. Standard Deviational Ellipse is a modified version of standard distance that captures the shape of this distribution by showing any directional bias in the pattern.

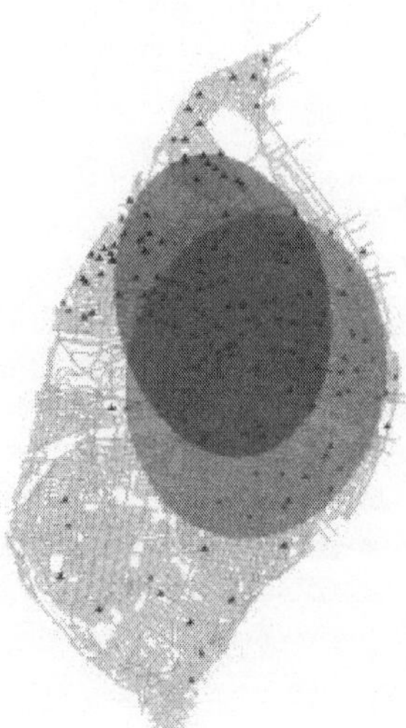

Other Statistical Operations: These are some of the many spatial statistics that can be treated as analogous to numerical data sets. Just as one can examine the median, or the distribution of a numerical data set, one can perform the same operations on spatial statistics. For example, a research question might be how does a particular frequency compare to another pattern, or another area? If Detroit averages one supermarket per square mile, how does that compare to New York's fifteen supermarkets per square mile? Once this measure of frequency is reduced to a number, the entire set of non-spatial statistical tests can be performed, such as T-test comparing the frequency of two study areas, or K-S tests for comparing frequency distributions.

Dispersion & Arrangement: Another set of questions in point pattern analysis concerns the relative pattern or arrangement of the points. Point patterns can be categorized as random, uniform, clustered or dispersed along the following two continuums:

- Random vs. Uniform (stratified, regular)
- Clustered vs. Dispersed

These two continuums are not necessarily related, and therefore a pattern of points could be randomly distributed in a clustered way (far right image), or stratified and dispersed (middle image). The attributes of one continuum have no effect on the attributes of the other continuum. The difference between these types of distributions can be easily seen in our example, but there are techniques to quantify the amount of stratification or clustering. This allows one to determine exactly how clustered something is, or to compare two sets of points. One can see applications to population science, for example being able to determine that the incidence of a disease is randomly distributed across a city or that it is clustered around a particular pollution source.

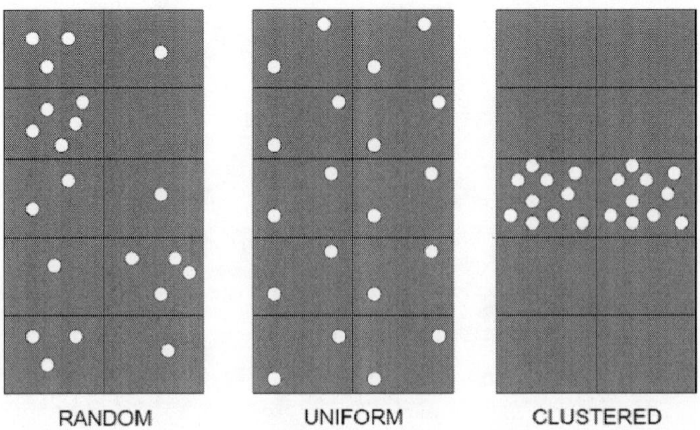

Complete Spatial Randomness: There are a number of techniques specifically designed for pattern analysis of point data. A concept that is common in these techniques is complete spatial randomness (CSR). CSR is a random or Poisson distribution of points in an area, the spatial version of a random or Poisson distribution of values in a numerical data set. Actual data is often contrasted with CSR and serves as a baseline for analysis.

Quadrat Analysis: In quadrat analysis you divide your study area into subsections of equal size, count the frequency of points in each subsection and then calculate the frequency of points in each subsection.

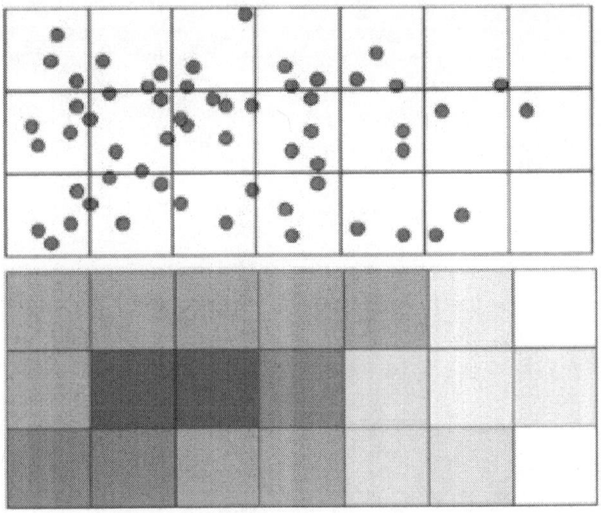

This can be performed for an entire data set, or can be conducted as a sampling technique to give one an idea about the total frequency or distribution of a larger area.

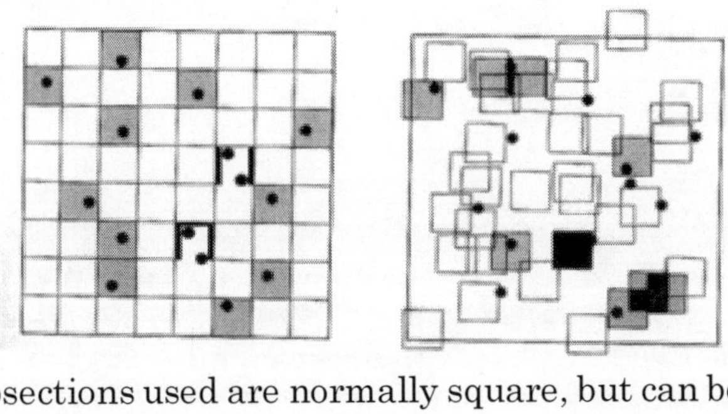

The subsections used are normally square, but can be of any shape depending on need. This method provides a study-specific, numerical measure of frequency. The different values for each cell also give a mean value of frequency of points across the study area, as well as the ability to calculate variance between subsections. These variance and mean frequency statistics are commonly combined to create a further statistic called variance to mean ratio (VMR). This is a unit-less statistic describing the spatial arrangement of your points. In general stratified distributions have a VMR of approximately 0, random distributions a VMR of 1 and clustered distributions a VMR of above 1.

Quadrat analysis is a simple, customizable, and valuable analysis tool. However, there are some criticisms of the method. Primarily, these arise from the ability to vary the size of the quadrat regions for each study. Therefore, quadrat analysis can provide different results for the same dataset using different sizes (a version of the Modifiable Areal Unit Problem). Also, the result of quadrat analysis is a single measure of frequency for the entire study, so variations within the regions are disguised. This method must be applied in an appropriate manner for each study and an understanding of the dataset is needed before performing this technique

Nearest Neighbour: In this analysis the distance of each point to its nearest neighbour is measured and the average nearest neighbour distance for all points is determined Figure depicted earlier.

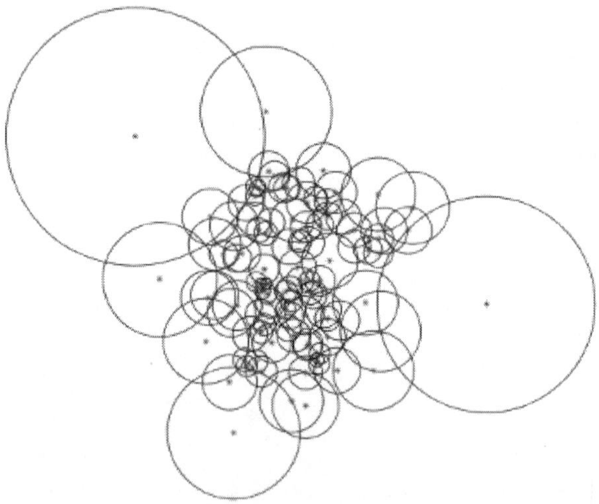

The results can be compared between two actual areas, or by comparing your study area to CSR. An index statistic called the Nearest Neighbour Index (NNI) can be created by dividing the observed average distance by the expected average distance (of CSR) to create a unit-less measure of dispersion. Generally, random patterns have a NNI value of about 1, clustered patterns 0, and dispersed patterns larger than 2. An application of this technique might be studying how far apart hospital emergency rooms were to each other.

These are some of the more common spatial analysis techniques used for point patterns. There are a number of refinements and extensions of these techniques, including Ripleys K Function, the G function, the F function, and Moran's I. Point pattern analysis, interpolation, and visualization are powerful techniques for analyzing population data, and are easy to add to your arsenal. View the Additional Reading section for more advanced work on this topic.

Related Concepts:

- Spatial Dependence: Defined by Michael Goodchild as "the propensity for nearby locations to influence each other and to possess similar attributes", spatial dependence is when spatial data are not independent from each other, violating an underlying assumption of many statistics. Statistical measures of the similarity of points are called spatial autocorrelation. For more one this topic, see: Spatial Analysis and GIS: Spatial Dependence at CSSIS.org.
- Edge Effects: Edge effects are a type of Boundary problem, an issue particular to spatial analysis. They are artifacts produced by the necessity of drawing a limit around your chosen study area. For more one this topic see: Boundary Problems at Wikipedia.org.
- First-order and second order effects in spatial analysis: First-order effects concern the way in which the expected value of the process varies across space, while second-order effects describe the correlation between values of the process at different regions in space. Region-wide trends are first-order effects while spatial dependence is a second-order effect.
- Stationary: An adjective in spatial analysis which means there is no first-order effects for the particular object being studied.

Chapter 6

Spatial Regression and Geostatistical Models

Kriging

In statistics, originally in geostatistics, Kriging or Gaussian process regression is a method of interpolation for which the interpolated values are modelled by a Gaussian process governed by prior covariances, as opposed to a piecewise-polynomial spline chosen to optimize smoothness of the fitted values.

Under suitable assumptions on the priors, Kriging gives the best linear unbiased prediction of the intermediate values. Interpolating methods based on other criteria such as smoothness need not yield the most likely intermediate values.

The method is widely used in the domain of spatial analysis and computer experiments. The technique is also known as Kolmogorov Wiener prediction.

The theoretical basis for the method was developed by the French mathematician Georges Matheron based on the Master's thesis of Danie G. Krige, the pioneering plotter of distance-weighted average gold grades at the Witwatersrand reef complex in South Africa.

Krige sought to estimate the most likely distribution of gold based on samples from a few boreholes.

The English verb is *to krige* and the most common noun is *kriging*; both are often pronounced with a hard "g", following the pronunciation of the name "Krige".

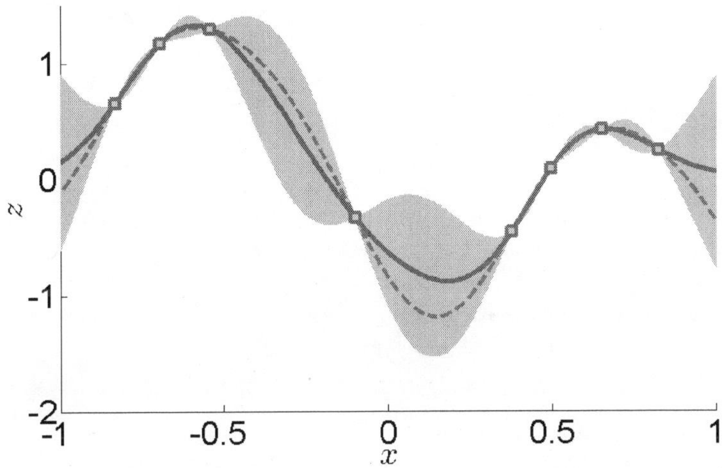

Figure: *Example of one-dimensional data interpolation by kriging, with confidence intervals. Squares indicate the location of the data. The Kriging interpolation, shown in red, runs along the means of the normally distributed confidence intervals shown in gray. The dashed curve shows a spline that while smooth nevertheless departs significantly from the expected intermediate values given by those means.*

Main Principles

Related Terms and Techniques

The basic idea of Kriging is to predict the value of a function at a given point by computing a weighted average of the known values of the function in the neighbourhood of the point. The method is mathematically closely related to regression analysis. Both theories derive a best linear unbiased estimator, based on assumptions on covariances, make use of Gauss-Markov theorem to prove independence of the estimate and error, and make use of very similar formulae. Even so, they are useful in different frameworks: Kriging is made for estimation of a single realisation of a random field, while regression models are based on multiple observations of a multivariate data set.

The Kriging estimation may also be seen as a spline in a reproducing kernel Hilbert space, with the reproducing kernel given by the covariance function. The difference with the classical Kriging approach is provided by the interpretation: while the spline is motivated by a minimum norm interpolation based on a Hilbert space structure, Kriging is motivated by an expected squared prediction error based on a stochastic model.

Kriging with *polynomial trend surfaces* is mathematically identical to generalized least squares polynomial curve fitting.

Kriging can also be understood as a form of Bayesian inference. Kriging starts with a prior distribution over functions. This prior takes the form of a Gaussian process: N samples from a function will be normally distributed, where the covariance between any two samples is the covariance function (or kernel) of the Gaussian process evaluated at the spatial location of two points. A set of values is then observed, each value associated with a spatial location. Now, a new value can be predicted at any new spatial location, by combining the Gaussian prior with a Gaussian likelihood function for each of the observed values. The resulting posterior distribution is also Gaussian, with a mean and covariance that can be simply computed from the observed values, their variance, and the kernel matrix derived from the prior.

Geostatistical Estimator

In geostatistical models, sampled data is interpreted as a result of a random process. The fact that these models incorporate uncertainty in their conceptualization doesn't mean that the phenomenon - the forest, the aquifer, the mineral deposit - has resulted from a random process, but solely allows to build a methodological basis for the spatial inference of quantities in unobserved locations and to the quantification of the uncertainty associated with the estimator.

A stochastic process is simply, in the context of this model, a way to approach the set of data collected from the samples. The first step in geostatistical modulation is the creation of a random process that best describes the set of experimental observed data.

A value spatially located at x_1 (generic denomination of a set of geographic coordinates) is interpreted as a realisation $z(x_1)$ of the random variable $Z(x_1)$. In the space A, where the set of samples is dispersed, exists realisations of the random variables $Z(x_1), Z(x_2), \cdots, Z(x_N)$, correlated between themselves.

The set of random variables, constitutes a random function of which only one realisation is known $z(x_i)$ - the set of experimental data. With only one realisation of each random variable it's theoretically impossible to determine any statistical parameter of the individual variables or the function.

The proposed solution in the geostatistical formalism consists in assuming various degrees of stationarity in the random function, in order to make possible the inference of some statistic values.

For instance, if a workgroup of scientists assumes appropriate, based on the homogeneity of samples in area A where the variable is distributed, the hypothesis that the first moment is stationary (i.e. all random variables have the same mean), then, they are implying that the mean can be estimated by the arithmetic mean of sampled values. Judging a hypothesis like this as appropriate is the same as considering that sample values are sufficiently homogeneous to validate that representativity.

The hypothesis of stationarity related to the second moment is defined in the following way: the correlation between two random variables solely depends on the spatial distance that separates them and is independent of its location:

$$C(Z(x_1), Z(x_2)) = C(Z(x_i), Z(x_i + \mathbf{h})) = C(\mathbf{h})$$

$$\gamma(Z(x_1), Z(x_2)) = \gamma(Z(x_i), Z(x_i + \mathbf{h})) = \gamma(\mathbf{h})$$

where $\mathbf{h} = (x_1, x_2) = (x_i, x_i + \mathbf{h})$

This hypothesis allows to infer those two measures - the variogram and the covariogram - based on the N samples:

$$\gamma(\mathbf{h}) = \frac{1}{2N(\mathbf{h})} \sum_{i=1}^{N(\mathbf{h})} \left(Z(x_i) - Z(x_i + \mathbf{h}) \right)^2$$

$$C(\mathbf{h}) = \frac{1}{N(\mathbf{h})} \sum_{i=1}^{N(\mathbf{h})} \left(Z(x_i) Z(x_i + \mathbf{h}) \right) - m(x_i) m(x_i + \mathbf{h})$$

where $m(x_i) = \frac{1}{N(\mathbf{h})} \sum_{i=1}^{N(\mathbf{h})} Z(x_i)$

Linear Estimation

Spatial inference, or estimation, of a quantity $Z : \mathbb{R}^n \to \mathbb{R}$, at an unobserved location x_0, is calculated from a linear combination of the observed values $z_i = Z(x_i)$ and weights :

$$\hat{Z}(x_0) = \begin{bmatrix} w_1 & w_2 & \cdots & w_N \end{bmatrix} \cdot \begin{bmatrix} z_1 \\ z_2 \\ \vdots \\ z_N \end{bmatrix} = \sum_{i=1}^{n} w_i(x_0) \times Z(x_i)$$

The weights w_i are intended to summarize two extremely important procedures in a spatial inference process:

- reflect the structural "proximity" of samples to the estimation location, x_0

- at the same time, w_i they should have a desegregation effect, in order to avoid bias caused by eventual sample *clusters*

When calculating the weights , there are two objectives in the geostatistical formalism: *unbias* and *minimal variance of estimation.*

If the cloud of real values $Z(x_0)$ is plotted against the estimated values $\hat{Z}(x_0)$, the criterion for global unbias, *intrinsic stationarity* or $(\hat{Z}(x) - Z(x))$ wide sense stationarity of the field, implies that the mean of the estimations must be equal to mean of the real values.

The second criterion says that the mean of the squared deviations must be minimal, which means that when the cloud of estimated values *versus* the cloud real values is more disperse, the estimator is more imprecise.

Methods

Depending on the stochastic properties of the random field and the various degrees of stationarity assumed, different methods for calculating the weights can be deducted, i.e. different types of kriging apply. Classical methods are:

- *Ordinary Kriging* assumes stationarity of the first moment of all random variables: $E\{Z(x_i)\} = E\{Z(x_0)\} = m$, where is unknown.
- *Simple Kriging* assumes a known stationary mean: $E\{Z(x)\} = m$, where is known.
- *Universal Kriging* assumes a general polynomial trend model, such as linear trend model.

$$E\{Z(x)\} = \sum_{k=0}^{p} \beta_k f_k(x)$$

- *IRFk-kriging* assumes $E\{Z(x)\}$ to be an unknown polynomial in x.
- *Indicator Kriging* uses indicator functions instead of the process itself, in order to estimate transition probabilities.
- *Multiple-indicator kriging* is a version of indicator kriging working with a family of indicators. However, MIK has fallen out of favour as an interpolation technique in recent years. This is due to some inherent difficulties related to operation and model validation. Conditional simulation is fast becoming the accepted replacement technique in this case.

- *Disjunctive Kriging* is a nonlinear generalisation of Kriging.
- *Lognormal Kriging* interpolates positive data by means of logarithms.

Ordinary Kriging

The unknown value $Z(x_0)$ is interpreted as a random variable located in x_0, as well as the values of neighbours samples $Z(x_i), i = 1, \cdots, N$. The estimator $\hat{Z}(x_0)$ is also interpreted as a random variable located in x_0, a result of the linear combination of variables.

In order to deduce the kriging system for the assumptions of the model, the following error committed while estimating $Z(x)$ in x_0 is declared:

$$\epsilon(x_0) = \hat{Z}(x_0) - Z(x_0) = \begin{bmatrix} W^T & -1 \end{bmatrix} \cdot \begin{bmatrix} Z(x_i) & \cdots & Z(x_N) & Z(x_0) \end{bmatrix}^T = \sum_{i=1}^{N} w_i(x_0) \times Z(x_i) - Z(x_0)$$

The two quality criteria referred to previously can now be expressed in terms of the mean and variance of the new random variable $\epsilon(x_0)$:

Lack of Bias

Since the random function is stationary, $E(Z(x_i)) = E(Z(x_0)) = m$, the following constraint is observed:

$$E\left(\epsilon(x_0)\right) = 0 \Leftrightarrow \sum_{i=1}^{N} w_i(x_0) \times E(Z(x_i)) - E(Z(x_0)) = 0 \Leftrightarrow$$

$$\Leftrightarrow m \sum_{i=1}^{N} w_i(x_0) - m = 0 \Leftrightarrow \sum_{i=1}^{N} w_i(x_0) = 1 \Leftrightarrow \mathbf{1}^T \cdot W = 1$$

In order to ensure that the model is unbiased, the weights must sum to one.

Minimum Variance

Two estimators can have $E\left[\epsilon(x_0)\right] = 0$, but the dispersion around their mean determines the difference between the quality of estimators. To find an estimator with minimum variance, we need to minimize $E\left(\epsilon(x_0)^2\right)$.

$$\begin{aligned} Var(\epsilon(x_0)) &= Var\left(\begin{bmatrix} W^T & -1 \end{bmatrix} \cdot \begin{bmatrix} Z(x_i) & \cdots & Z(x_N) & Z(x_0) \end{bmatrix}^T\right) = \\ &\overset{*}{=} \begin{bmatrix} W^T & -1 \end{bmatrix} \cdot Var\left(\begin{bmatrix} Z(x_i) & \cdots & Z(x_N) & Z(x_0) \end{bmatrix}^T\right) \cdot \begin{bmatrix} W \\ -1 \end{bmatrix} \end{aligned}$$

$$Var(\epsilon(x_0)) \overset{*}{=} \begin{bmatrix} W^T & -1 \end{bmatrix} \cdot \begin{bmatrix} Var_{x_i} & Cov_{x_i x_0} \\ Cov^T_{x_i x_0} & Var_{x_0} \end{bmatrix} \cdot \begin{bmatrix} W \\ -1 \end{bmatrix}$$

* where the literals $\left\{ Var_{x_i}, Var_{x_0}, Cov_{x_i x_0} \right\}$ stand for

$$\left\{ Var\left(\begin{bmatrix} Z(x_1) & \cdots & Z(x_N) \end{bmatrix}^T\right), Var(Z(x_0)), Cov\left(\begin{bmatrix} Z(x_1) & \cdots & Z(x_N) \end{bmatrix}^T, Z(x_0)\right) \right\}.$$

Once defined the covariance model $C(\mathbf{h})$ or variogram, or $\gamma(\mathbf{h})$, valid in all field of analysis of $Z(x)$, than we can write an expression for the estimation variance of any estimator in function of the covariance between the samples and the covariances between the samples and the point to estimate:

$$\begin{cases} Var(\epsilon(x_0)) = W^T \cdot Var_{x_i} \cdot W - Cov^T_{x_i x_0} \cdot W - W^T \cdot Cov_{x_i x_0} + Var_{x_0} \\ Var(\epsilon(x_0)) = Cov(0) + \sum_i \sum_j w_i w_j Cov(x_i, x_j) - 2 \sum_i w_i C(x_i, x_0) \end{cases}$$

Some conclusions can be asserted from this expression. The variance of estimation:

- is not quantifiable to any linear estimator, once the stationarity of the mean and of the spatial covariances, or variograms, are assumed.
- grows with the covariance between samples $C(x_i, x_j)$, i.e. to the same distance to the estimating point, if the samples are proximal to each other, then the *clustering effect*, or informational redundancy, is bigger, so the estimation is worst. This conclusion is valid to any value of the weights: a preferential grouping of samples is always worst, which means that for the same number of samples the estimation variance grows with the relative weight of the sample *clusters*.
- grows when the covariance between the samples and the point to estimate decreases. This means that, when the samples are farther away from x_0, the worse the estimation.
- grows with the *a priori* variance $C(0)$ of the variable $Z(x)$. When the variable is less disperse, the variance is lower in any point of the area A.
- does not depend on the values of the samples. This means that the same spatial configuration (with the same geometrical relations between samples and the point to estimate) always reproduces the same estimation variance in any part of the area

A. This way, the variance does not measures the uncertainty of estimation produced by the local variable.

Kriging System

$$\underset{W}{\text{minimize}} \quad W^T \cdot Var_{x_i} \cdot W - Cov^T_{x_i x_0} \cdot W - W^T \cdot Cov_{x_i x_0} + Var_{x_0}$$

$$\text{subject to} \quad \mathbf{1}^T \cdot W = 1$$

Solving this optimization problem results in the *kriging system*:

$$\begin{bmatrix} \hat{W} \\ \mu \end{bmatrix} = \begin{bmatrix} Var_{x_i} & \mathbf{1} \\ \mathbf{1}^T & 0 \end{bmatrix}^{-1} \cdot \begin{bmatrix} Cov_{x_i x_0} \\ 1 \end{bmatrix} = \begin{bmatrix} \gamma(x_1,x_1) & \cdots & \gamma(x_1,x_n) & 1 \\ \vdots & \ddots & \vdots & \vdots \\ \gamma(x_n,x_1) & \cdots & \gamma(x_n,x_n) & 1 \\ 1 & \cdots & 1 & 0 \end{bmatrix}^{-1} \begin{bmatrix} \gamma(x_1,x^*) \\ \vdots \\ \gamma(x_n,x^*) \\ 1 \end{bmatrix}$$

the additional parameter μ is a Lagrange multiplier used in the minimization of the kriging error $\sigma_k^2(x)$ to honour the unbiasedness condition.

Simple Kriging

Simple kriging is mathematically the simplest, but the least general. It assumes the expectation of the random field to be known, and relies on a covariance function. However, in most applications neither the expectation nor the covariance are known beforehand.

The practical assumptions for the application of *simple kriging* are:

- wide sense stationarity of the field.
- The expectation is zero everywhere: $\mu(x) = 0$.
- Known covariance function $c(x, y) = \text{Cov}(Z(x), Z(y))$

Kriging System

The *kriging weights* of *simple kriging* have no unbiasedness condition and are given by the *simple kriging equation system*:

$$\begin{pmatrix} w_1 \\ \vdots \\ w_n \end{pmatrix} = \begin{pmatrix} c(x_1,x_1) & \cdots & c(x_1,x_n) \\ \vdots & \ddots & \vdots \\ c(x_n,x_1) & \cdots & c(x_n,x_n) \end{pmatrix}^{-1} \begin{pmatrix} c(x_1,x_0) \\ \vdots \\ c(x_n,x_0) \end{pmatrix}$$

This is analogous to a linear regression of $Z(x_0)$ on the other .

Estimation

The interpolation by simple kriging is given by:

$$\hat{Z}(x_0) = \begin{pmatrix} z_1 \\ \vdots \\ z_n \end{pmatrix}' \begin{pmatrix} c(x_1,x_1) & \cdots & c(x_1,x_n) \\ \vdots & \ddots & \vdots \\ c(x_n,x_1) & \cdots & c(x_n,x_n) \end{pmatrix}^{-1} \begin{pmatrix} c(x_1,x_0) \\ \vdots \\ c(x_n,x_0) \end{pmatrix}$$

The kriging error is given by:

$$\mathrm{Var}\left(\hat{Z}(x_0) - Z(x_0)\right) = \underbrace{c(x_0,x_0)}_{\mathrm{Var}(Z(x_0))} - \underbrace{\begin{pmatrix} c(x_1,x_0) \\ \vdots \\ c(x_n,x_0) \end{pmatrix}' \begin{pmatrix} c(x_1,x_1) & \cdots & c(x_1,x_n) \\ \vdots & \ddots & \vdots \\ c(x_n,x_1) & \cdots & c(x_n,x_n) \end{pmatrix}^{-1} \begin{pmatrix} c(x_1,x_0) \\ \vdots \\ c(x_n,x_0) \end{pmatrix}}_{\mathrm{Var}(\hat{Z}(x_0))}$$

which leads to the generalised least squares version of the Gauss-Markov theorem:

$$\mathrm{Var}(Z(x_0)) = \mathrm{Var}(\hat{Z}(x_0)) + \mathrm{Var}\left(\hat{Z}(x_0) - Z(x_0)\right).$$

Properties

- The kriging estimation is unbiased: $E[\hat{Z}(x_i)] = E[Z(x_i)]$
- The kriging estimation honours the actually observed value: $\hat{Z}(x_i) = Z(x_i)$ (assuming no measurement error is incurred)
- The kriging estimation $\hat{Z}(x)$ is the best linear unbiased estimator of $Z(x)$ if the assumptions hold. However (e.g. Cressie 1993):
  - o As with any method: If the assumptions do not hold, kriging might be bad.
  - o There might be better nonlinear and/or biased methods.
  - o No properties are guaranteed, when the wrong variogram is used. However typically still a 'good' interpolation is achieved.
  - o Best is not necessarily good: e.g. In case of no spatial dependence the kriging interpolation is only as good as the arithmetic mean.
- Kriging provides σ_k^2 as a measure of precision. However this measure relies on the correctness of the variogram.

Applications

Although kriging was developed originally for applications in geostatistics, it is a general method of statistical interpolation that can be applied within any discipline to sampled data from random fields that satisfy the appropriate mathematical assumptions.

To date kriging has been used in a variety of disciplines, including the following:

- Environmental science
- Hydrogeology
- Mining
- Natural resources
- Remote sensing
- Real estate appraisal
- Integrated Circuit Analysis and Optimization

Design and Analysis of Computer Experiments

Another very important and rapidly growing field of application, in engineering, is the interpolation of data coming out as response variables of deterministic computer simulations, e.g. finite element method (FEM) simulations. In this case, kriging is used as a metamodelling tool, i.e. a black box model built over a designed set of computer experiments. In many practical engineering problems, such as the design of a metal forming process, a single FEM simulation might be several hours or even a few days long. It is therefore more efficient to design and run a limited number of computer simulations, and then use a kriging interpolator to rapidly predict the response in any other design point. Kriging is therefore used very often as a so-called surrogate model, implemented inside optimization routines.

Lagrange Multipliers

In mathematical optimization, the method of Lagrange multipliers (named after Joseph Louis Lagrange) is a strategy for finding the local maxima and minima of a function subject to equality constraints.

For instance, consider the optimization problem

maximize $f(x, y)$

subject to $g(x, y) = c$.

We need both f and g to have continuous first partial derivatives. We introduce a new variable (λ) called a Lagrange multiplier and study the Lagrange function (or Lagrangian) defined by

$$\Lambda(x, y, \lambda) = f(x, y) + \lambda \cdot \big(g(x, y) - c\big),$$

where the λ term may be either added or subtracted. If $f(x_0, y_0)$ is a maximum of $f(x, y)$ for the original constrained problem, then there exists λ_0 such that (x_0, y_0, λ_0) is a stationary point for the Lagrange function (stationary points are those points where the partial derivatives of Λ are zero). However, not all stationary points yield a solution of the original problem. Thus, the method of Lagrange multipliers yields a necessary condition for optimality in constrained problems. Sufficient conditions for a minimum or maximum also exist.

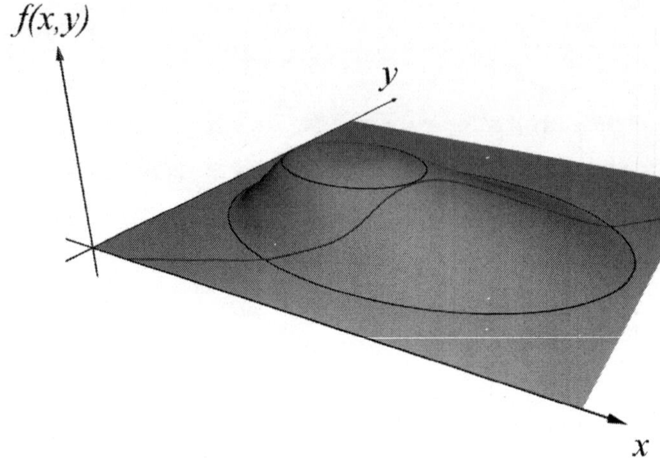

Figure: *Find x and y to maximize f(x, y) subject to a constraint (shown in red) g(x, y) = c.*

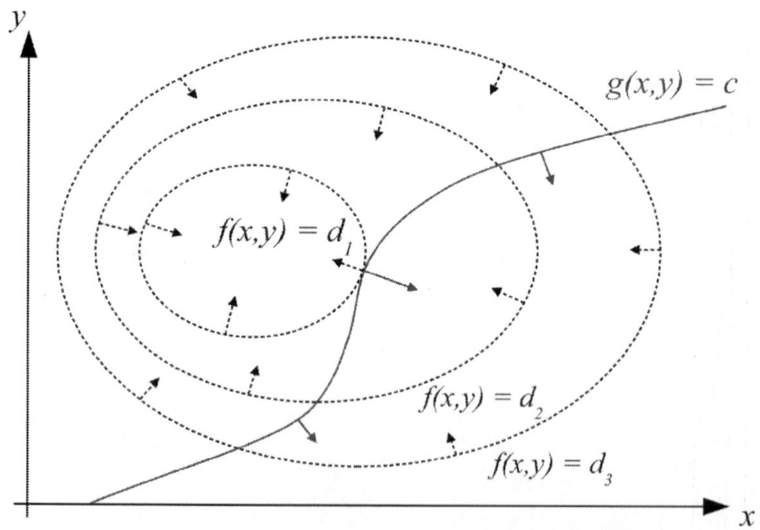

Figure: *The red line shows the constraint g(x, y) = c. The blue lines are contours of f(x, y). The point where the red line tangentially touches a blue contour is our solution. Since $d_1 > d_2$, the solution is a maximization of f(x, y).*

One of the most common problems in calculus is that of finding maxima or minima (in general, "extrema") of a function, but it is often difficult to find a closed form for the function being extremized. Such difficulties often arise when one wishes to maximize or minimize a function subject to fixed outside conditions or constraints. The method of Lagrange multipliers is a powerful tool for solving this class of problems without the need to explicitly solve the conditions and use them to eliminate extra variables.

Consider the two-dimensional problem introduced above:

maximize $f(x, y)$

subject to $g(x, y) = c$

Lagrange multipliers relies on the intuition that at a maximum $f(x, y)$ cannot be increasing in the direction of any neighbouring point where $g = c$. If it were, we could walk along $g = c$ to get higher, meaning that the starting point wasn't actually the maximum.

We can visualize contours of f given by $f(x, y) = d$ for various values of d, and the contour of g given by $g(x, y) = c$.

Suppose we walk along the contour line with $g = c$. We are interested in finding points where f does not change as we walk, since these points might be maxima. There are two ways this could happen: First, we could be following a contour line of f, since by definition f does not change as we walk along its contour lines. This would mean that the contour lines of f and g are parallel here. The second possibility is that we have reached a "level" part of f, meaning that f does not change in any direction.

To check the first possibility, notice that since the gradient of a function is perpendicular to the contour lines, the contour lines of f and g are parallel if and only if the gradients of f and g are parallel. Thus we want points (x, y) where $g(x, y) = c$ and

$$\nabla_{x,y} f = -\lambda \nabla_{x,y} g,$$

for some λ

where

$$\nabla_{x,y} f = \left(\frac{\partial f}{\partial x}, \frac{\partial f}{\partial y}\right), \qquad \nabla_{x,y} g = \left(\frac{\partial g}{\partial x}, \frac{\partial g}{\partial y}\right).$$

are the respective gradients. The constant λ is required because although the two gradient vectors are parallel, the magnitudes of the gradient vectors are generally not equal. (The negative is traditional). This constant is called the Lagrange multiplier.

Notice that this method also solves the second possibility: if f is level, then its gradient is zero, and setting $\lambda = 0$ is a solution regardless of g.

To incorporate these conditions into one equation, we introduce an auxiliary function

$$\Lambda(x, y, \lambda) = f(x, y) + \lambda \cdot \big(g(x, y) - c\big),$$

and solve

$$\nabla_{x,y,\lambda} \Lambda(x, y, \lambda) = 0.$$

This is the method of Lagrange multipliers. Note that $\nabla_{\lambda} \Lambda(x, y, \lambda) = 0$ implies $g(x, y) = c$.

The constrained extrema of f are *critical points* of the Lagrangian Λ, but they are not *local extrema* of $\wedge$.

One may reformulate the Lagrangian as a Hamiltonian, in which case the solutions are local minima for the Hamiltonian. This is done in optimal control theory, in the form of Pontryagin's minimum principle.

The fact that solutions of the Lagrangian are not necessarily extrema also poses difficulties for numerical optimization. This can be addressed by computing the *magnitude* of the gradient, as the zeros of the magnitude are necessarily local minima, as illustrated in the numerical optimization example.

Handling Multiple Constraints

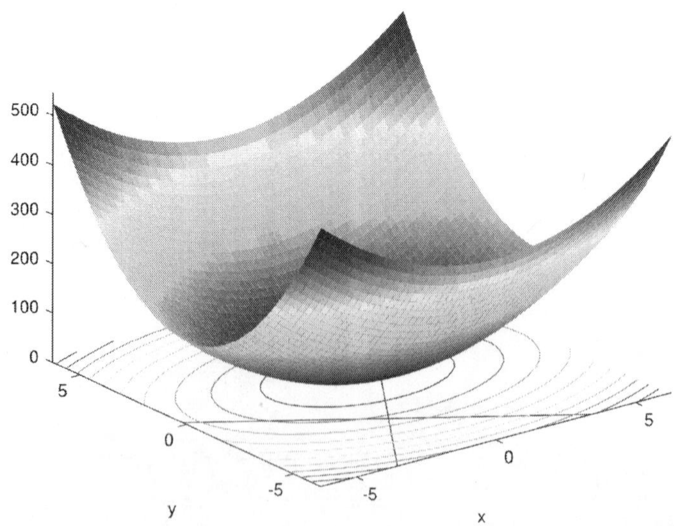

Figure: *A paraboloid, some of its level sets (or contour lines) and 2 line constraints.*

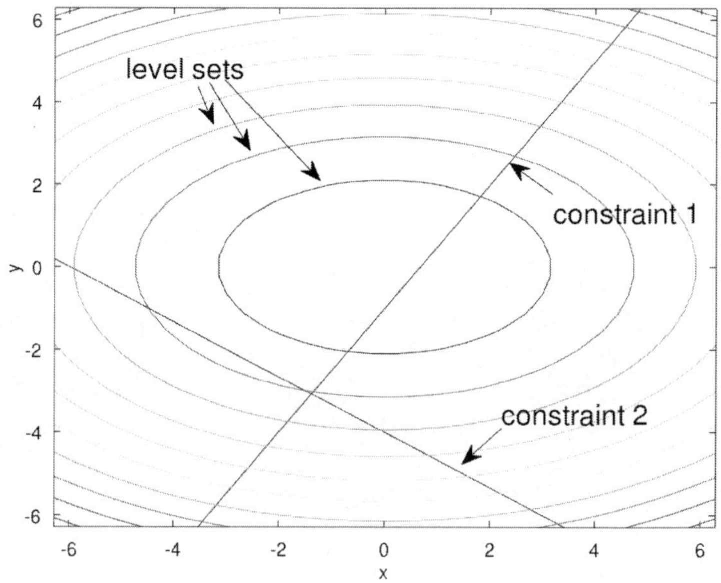

Figure: *Zooming in on the levels sets and constraints, we see that the two constraint lines intersect to form a "joint" constraint that is a point. Since there is only one point to analyze, the corresponding point on the paraboloid is automatically a minimum and maximum. Yet the simplified reasoning presented in sections above seems to fail because the level set definitely appears to "cross" the point and at the same time its gradient is not parallel to the gradients of either constraint. This shows we must refine our explanation of the method to handle the kinds of constraints that are formed when we have more than one constraint acting at once.*

The method of *Lagrange multipliers* can also accommodate multiple constraints. To see how this is done, we need to re-examine the problem in a slightly different manner because the concept of "crossing" discussed above becomes rapidly unclear when we consider the types of constraints that are created when we have more than one constraint acting together.

As an example, consider a paraboloid with a constraint that is a single point (as might be created if we had 2 line constraints that intersect). The level set (i.e., contour line) clearly appears to "cross" that point and its gradient is clearly not parallel to the gradients of either of the two line constraints. Yet, it is obviously a maximum *and* a minimum because there is only one point on the paraboloid that meets the constraint.

While this example seems a bit odd, it is easy to understand and is representative of the sort of "effective" constraint that appears quite often when we deal with multiple constraints intersecting. Thus, we take a slightly different approach below to explain and derive the Lagrange Multipliers method with any number of constraints.

Throughout this section, the independent variables will be denoted by $x_1, x_2, ..., x_N$ and, as a group, we will denote them as $p = (x_1, x_2, ..., x_N)$. Also, the function being analyzed will be denoted by $f(p)$ and the constraints will be represented by the equations $g_1(p) = g_2(p) = ... = g_M(p) = 0$.

The basic idea remains essentially the same: if we consider only the points that satisfy the constraints (i.e., are *in* the constraints), then a point $(p, f(p))$ is a stationary point (i.e., a point in a "flat" region) of f if and only if the constraints at that point do not allow movement in a direction where f changes value.

Once we have located the stationary points, we need to do further tests to see if we have found a minimum, a maximum or just a stationary point that is neither.

We start by considering the level set of f at $(p, f(p))$. The set of vectors $\{v_L\}$ containing the directions in which we can move and still remain in the same level set are the directions where the value of f does not change (i.e., the change equals zero). Thus, for every vector v in $\{v_L\}$, the following relation must hold:

$$\frac{\partial f}{\partial x_1} v_{x_1} + \frac{\partial f}{\partial x_2} v_{x_2} + \cdots + \frac{\partial f}{\partial x_N} v_{x_N} = 0$$

where the notation v_{xK} above means the x_K-component of the vector v. The equation above can be rewritten in a more compact geometric form that helps our intuition:

$$\underbrace{\begin{bmatrix} \frac{\partial f}{\partial x_1} & \frac{\partial f}{\partial x_2} & \cdots & \frac{\partial f}{\partial x_N} \end{bmatrix}}_{\nabla f^T} \underbrace{\begin{bmatrix} v_{x_1} \\ v_{x_2} \\ \vdots \\ v_{x_N} \end{bmatrix}}_{v} = 0$$

is the same as writing

$$\nabla f^T \cdot v = 0.$$

This makes it clear that if we are at p, then *all* directions from this point that do *not* change the value of f *must be perpendicular* to "$f(p)$ (the gradient of f at p).

Now let us consider the effect of the constraints. Each constraint limits the directions that we can move from a particular point and still satisfy the constraint. We can use the same procedure, to look for the set of vectors $\{v_C\}$ containing the directions in which we can move and still satisfy the constraint. As above, for every vector v in $\{v_C\}$, the following relation must hold:

$$\Delta g = \frac{\partial g}{\partial x_1} v_{x_1} + \frac{\partial g}{\partial x_2} v_{x_2} + \cdots + \frac{\partial g}{\partial x_N} v_{x_N} = 0 \quad \Rightarrow \quad \nabla g^T \cdot v = 0.$$

From this, we see that at point *p*, all directions from this point that will still satisfy this constraint must be perpendicular to "*g*(*p*).

Now we are ready to refine our idea further and complete the method: *a point on f is a constrained stationary point if and only if the direction that changes f violates at least one of the constraints, i.e., has no "component" in the "legal" space perpendicular to* $\nabla g(p)$. (We can see that this is true because if a direction that changes *f* did *not* violate any constraints, then there would be a "legal" point nearby with a higher or lower value for *f* and the current point would then not be a stationary point.) Mathematically, this means that the gradient of *f* at this constrained stationary point is perpendicular to the space spanned by the set of vectors $\{v_C\}$, which in turn is perpendicular to the gradients of the constraints *g*.

Single Constraint Revisited

For a single constraint, we use the statement above to say that at stationary points the direction that changes *f* is in the same direction that violates the constraint. To determine if two vectors are in the same direction, we note that if two vectors start from the same point and are "in the same direction", then one vector can always "reach" the other by changing its length and/or flipping to point the opposite way along the same direction line. In this way, we can succinctly state that two vectors point in the same direction if and only if one of them can be multiplied by some real number such that they become equal to the other. So, for our purposes, we require that:

$$\nabla f(p) = \lambda \nabla g(p) \qquad \Rightarrow \qquad \nabla f(p) - \lambda \nabla g(p) = 0.$$

If we now add another simultaneous equation to guarantee that we only perform this test when we are at a point that satisfies the constraint, we end up with 2 simultaneous equations that when solved, identify all constrained stationary points:

$$\begin{cases} g(p) = 0 & p \text{ satisfies constraint} \\ \nabla f(p) - \lambda \nabla g(p) = 0 & p \text{ is a stationary point.} \end{cases}$$

Note that the above is a succinct way of writing the equations. Fully expanded, there are $N+1$ simultaneous equations that need to be solved for the $N+1$ variables which are λ and $x_1, x_2, ..., x_N$:

$$\begin{aligned}
&g(x_1,x_2,\dots,x_N) &&= 0\\
&\frac{\partial f}{\partial x_1}(x_1,x_2,\dots,x_N)-\lambda\frac{\partial g}{\partial x_1}(x_1,x_2,\dots,x_N) &&= 0\\
&\frac{\partial f}{\partial x_2}(x_1,x_2,\dots,x_N)-\lambda\frac{\partial g}{\partial x_2}(x_1,x_2,\dots,x_N) &&= 0\\
& &&\vdots\\
&\frac{\partial f}{\partial x_N}(x_1,x_2,\dots x_N)-\lambda\frac{\partial g}{\partial x_N}(x_1,x_2,\dots,x_N) &&= 0.
\end{aligned}$$

Multiple Constraints

For more than one constraint, similar reasoning applies. Each constraint function, g, has a space of allowable directions at p: the space of vectors perpendicular to $\nabla g(p)$. The set of directions that are allowed by all constraints is thus the space of directions perpendicular to all of the constraint gradients. Denote this space of allowable moves by A and denote the span of the constraint gradients by S. By the discussion above, $A = S^{\perp}$, the space of vectors perpendicular to every element of S.

If p is an optimum then any element not perpendicular to $\nabla f(p)$ is not an allowable direction. One can show that this implies $\nabla f(p) \in A^{\perp} = S$. Thus there are scalars $\lambda_1, \lambda_2, ..., \lambda_M$ such that

$$\nabla f(p)=\sum_{k=1}^{M}\lambda_k\nabla g_k(p) \quad\Rightarrow\quad \nabla f(p)-\sum_{k=1}^{M}\lambda_k\nabla g_k(p)=0.$$

As before, we now add simultaneous equation to guarantee that we only perform this test when we are at a point that satisfies every constraint, we end up with simultaneous equations that when solved, identify all constrained stationary points:

$$\begin{cases} g_1(p)=g_2(p)=\cdots=g_M(p)=0 & p \text{ satisfies all constraints}\\ \nabla f(p)-\sum_{k=1}^{M}\lambda_k\nabla g_k(p)=0 & p \text{ is a stationary point.}\end{cases}$$

The method is complete now (from the standpoint of solving the problem of finding stationary points) but as mathematicians delight in doing, these equations can be further condensed into an even more elegant and succinct form. Lagrange must have cleverly noticed that the equations above look like partial derivatives of some larger scalar function L that takes all the $x_1, x_2, ..., x_N$ and all the $\lambda_1, \lambda_2, ..., \lambda_M$ as inputs. Next, he might then have noticed that setting every equation equal to zero is exactly what one would have to do to solve for the *unconstrained* stationary points of that larger function. Finally, he showed that a larger function L with partial derivatives that are exactly the ones we require can be constructed very simply as below:

$$L(x_1, x_2, \ldots, x_N, \lambda_1, \lambda_2, \ldots, \lambda_M) = f(x_1, x_2, \ldots, x_N) - \sum_{k=1}^{M} \lambda_k g_k(x_1, x_2, \ldots, x_N).$$

Solving the equation above for its *unconstrained* stationary points generates exactly the same stationary points as solving for the *constrained* stationary points of f under the constraints $g_1, g_2, ..., g_M$.

In Lagrange's honour, the function above is called a *Lagrangian*, the scalars $\lambda_1, \lambda_2, ..., \lambda_M$ are called *Lagrange Multipliers* and this optimization method itself is called *The Method of Lagrange Multipliers*.

The method of Lagrange multipliers is generalized by the Karush–Kuhn–Tucker conditions, which can also take into account inequality constraints of the form $h(\mathrm{x}) \le c$.

Interpretation of the Lagrange Multipliers

Often the Lagrange multipliers have an interpretation as some quantity of interest. For example, if the Lagrangian expression is

$$L(x_1, x_2, \ldots; \lambda_1, \lambda_2, \ldots) = f(x_1, x_2, \ldots) + \lambda_1(c_1 - g_1(x_1, x_2, \ldots)) + \lambda_2(c_2 - g_2(x_1, x_2, \ldots)) + \ldots$$

then

$$\frac{\partial L}{\partial c_k} = \lambda_k.$$

So, λ_k is the rate of change of the quantity being optimized as a function of the constraint variable. As examples, in Lagrangian mechanics the equations of motion are derived by finding stationary points of the action, the time integral of the difference between kinetic and potential energy. Thus, the force on a particle due to a scalar potential, $F = -\nabla V$, can be interpreted as a Lagrange multiplier determining the change in action (transfer of potential to kinetic energy) following a variation in the particle's constrained trajectory. In control theory this is formulated instead as costate equations.

Moreover, by the envelope theorem the optimal value of a Lagrange multiplier has an interpretation as the marginal effect of the corresponding constraint constant upon the optimal attainable value of the original objective function: if we denote values at the optimum with an asterisk, then it can be shown that

$$\frac{\mathrm{d}f(x_1^*(c_1,c_2,\ldots),x_2^*(c_1,c_2,\ldots),\ldots)}{\mathrm{d}c_k}=\lambda_k^*.$$

For example, in economics the optimal profit to a player is calculated subject to a constrained space of actions, where a Lagrange multiplier is the change in the optimal value of the objective function (profit) due to the relaxation of a given constraint (e.g. through a change in income); in such a context λ^* is the marginal cost of the constraint, and is referred to as the shadow price.

Sufficient Conditions

Sufficient conditions for a constrained local maximum or minimum can be stated in terms of a sequence of principal minors (determinants of upper-left-justified sub-matrices) of the bordered Hessian matrix of second derivatives of the Lagrangian expression.

Chapter 7

Statistical Inference for Spatial Data

In statistics, statistical inference is the process of deducing properties of an underlying distribution by analysis of data. Inferential statistical analysis infers properties about a population: this includes testing hypotheses and deriving estimates. The population is assumed to be larger than the observed data set; in other words, the observed data is assumed to be sampled from a larger population.

Inferential statistics can be contrasted with descriptive statistics. Descriptive statistics is solely concerned with properties of the observed data, and does not assume that the data came from a larger population.

Statistical inference makes propositions about a population, using data drawn from the population via some form of sampling. Given a hypothesis about a population, for which we wish to draw inferences, statistical inference consists of (firstly) selecting a statistical model of the process that generates the data and (secondly) deducing propositions from the model.

The conclusion of a statistical inference is a statistical proposition. Some common forms of statistical proposition are the following:

- an estimate, i.e. a particular value that best approximates some parameter of interest;
- a confidence interval (or set estimate), i.e. an interval constructed using a dataset drawn from a population so that, under repeated sampling of such datasets, such intervals would contain the true parameter value with the probability at the stated confidence level;

- a credible interval, i.e. a set of values containing, for example, 95% of posterior belief;
- rejection of a hypothesis;
- clustering or classification of data points into groups.

Models and Assumptions

Any statistical inference requires some assumptions. A statistical model is a set of assumptions concerning the generation of the observed data and similar data. Descriptions of statistical models usually emphasize the role of population quantities of interest, about which we wish to draw inference. Descriptive statistics are typically used as a preliminary step before more formal inferences are drawn.

Degree of Models/Assumptions

Statisticians distinguish between three levels of modelling assumptions;

- Fully parametric: The probability distributions describing the data-generation process are assumed to be fully described by a family of probability distributions involving only a finite number of unknown parameters. For example, one may assume that the distribution of population values is truly Normal, with unknown mean and variance, and that datasets are generated by 'simple' random sampling. The family of generalized linear models is a widely used and flexible class of parametric models.
- Non-parametric: The assumptions made about the process generating the data are much less than in parametric statistics and may be minimal. For example, every continuous probability distribution has a median, which may be estimated using the sample median or the Hodges–Lehmann–Sen estimator, which has good properties when the data arise from simple random sampling.
- Semi-parametric: This term typically implies assumptions 'in between' fully and non-parametric approaches. For example, one may assume that a population distribution has a finite mean. Furthermore, one may assume that the mean response level in the population depends in a truly linear manner on some covariate (a parametric assumption) but not make any parametric assumption describing the variance around that mean (i.e., about the presence or possible form of any heteroscedasticity). More generally, semi-parametric models can often be separated into 'structural' and 'random variation'

components. One component is treated parametrically and the other non-parametrically. The well-known Cox model is a set of semi-parametric assumptions.

Importance of Valid Models/Assumptions

Whatever level of assumption is made, correctly calibrated inference in general requires these assumptions to be correct; i.e., that the data-generating mechanisms really have been correctly specified.

Incorrect assumptions of 'simple' random sampling can invalidate statistical inference. More complex semi- and fully parametric assumptions are also cause for concern. For example, incorrectly assuming the Cox model can in some cases lead to faulty conclusions. Incorrect assumptions of Normality in the population also invalidates some forms of regression-based inference. The use of any parametric model is viewed skeptically by most experts in sampling human populations: "most sampling statisticians, when they deal with confidence intervals at all, limit themselves to statements about [estimators] based on very large samples, where the central limit theorem ensures that these [estimators] will have distributions that are nearly normal." In particular, a normal distribution "would be a totally unrealistic and catastrophically unwise assumption to make if we were dealing with any kind of economic population." Here, the central limit theorem states that the distribution of the sample mean "for very large samples" is approximately normally distributed, if the distribution is not heavy tailed.

Approximate Distributions

Given the difficulty in specifying exact distributions of sample statistics, many methods have been developed for approximating these.

With finite samples, approximation results measure how close a limiting distribution approaches the statistic's sample distribution: For example, with 10,000 independent samples the normal distribution approximates (to two digits of accuracy) the distribution of the sample mean for many population distributions, by the Berry–Esseen theorem. Yet for many practical purposes, the normal approximation provides a good approximation to the sample-mean's distribution when there are 10 (or more) independent samples, according to simulation studies and statisticians' experience. Following Kolmogorov's work in the 1950s, advanced statistics uses approximation theory and functional analysis to quantify the error of approximation. In this approach, the metric geometry of probability distributions is studied; this approach quantifies approximation error with, for example, the Kullback–Leibler divergence, Bregman divergence, and the Hellinger distance.

With indefinitely large samples, limiting results like the central limit theorem describe the sample statistic's limiting distribution, if one exists. Limiting results are not statements about finite samples, and indeed are irrelevant to finite samples. However, the asymptotic theory of limiting distributions is often invoked for work with finite samples. For example, limiting results are often invoked to justify the generalized method of moments and the use of generalized estimating equations, which are popular in econometrics and biostatistics. The magnitude of the difference between the limiting distribution and the true distribution (formally, the 'error' of the approximation) can be assessed using simulation. The heuristic application of limiting results to finite samples is common practice in many applications, especially with low-dimensional models with log-concave likelihoods (such as with one-parameter exponential families).

Randomization-based Models

For a given dataset that was produced by a randomization design, the randomization distribution of a statistic (under the null-hypothesis) is defined by evaluating the test statistic for all of the plans that could have been generated by the randomization design. In frequentist inference, randomization allows inferences to be based on the randomization distribution rather than a subjective model, and this is important especially in survey sampling and design of experiments. Statistical inference from randomized studies is also more straightforward than many other situations. In Bayesian inference, randomization is also of importance: in survey sampling, use of sampling without replacement ensures the exchangeability of the sample with the population; in randomized experiments, randomization warrants a missing at random assumption for covariate information.

Objective randomization allows properly inductive procedures. Many statisticians prefer randomization-based analysis of data that was generated by well-defined randomization procedures. (However, it is true that in fields of science with developed theoretical knowledge and experimental control, randomized experiments may increase the costs of experimentation without improving the quality of inferences.) Similarly, results from randomized experiments are recommended by leading statistical authorities as allowing inferences with greater reliability than do observational studies of the same phenomena. However, a good observational study may be better than a bad randomized experiment.

The statistical analysis of a randomized experiment may be based on the randomization scheme stated in the experimental protocol and does not need a subjective model.

However, at any time, some hypotheses cannot be tested using objective statistical models, which accurately describe randomized experiments or random samples. In some cases, such randomized studies are uneconomical or unethical.

Model-based Analysis of Randomized Experiments

It is standard practice to refer to a statistical model, often a linear model, when analyzing data from randomized experiments. However, the randomization scheme guides the choice of a statistical model. It is not possible to choose an appropriate model without knowing the randomization scheme. Seriously misleading results can be obtained analyzing data from randomized experiments while ignoring the experimental protocol; common mistakes include forgetting the blocking used in an experiment and confusing repeated measurements on the same experimental unit with independent replicates of the treatment applied to different experimental units.

Modes of Inference

Different schools of statistical inference have become established. These schools (or 'paradigms') are not mutually exclusive, and methods which work well under one paradigm often have attractive interpretations under other paradigms. The two main paradigms in use are frequentist and Bayesian inference, which are both summarized below.

Frequentist Inference

This paradigm calibrates the production of propositions by considering (notional) repeated sampling of datasets similar to the one at hand. By considering its characteristics under repeated sample, the frequentist properties of any statistical inference procedure can be described — although in practice this quantification may be challenging.

Examples of Frequentist Inference

- p-value
- Confidence interval

Frequentist Inference, Objectivity and Decision Theory

One interpretation of frequentist inference (or classical inference) is that it is applicable only in terms of frequency probability; that is, in terms of repeated sampling from a population. However, the approach of Neyman develops these procedures in terms of pre-experiment probabilities.

That is, before undertaking an experiment, one decides on a rule for coming to a conclusion such that the probability of being correct is controlled in a suitable way: such a probability need not have a frequentist or repeated sampling interpretation. In contrast, Bayesian inference works in terms of conditional probabilities (i.e., probabilities conditional on the observed data), compared to the marginal (but conditioned on unknown parameters) probabilities used in the frequentist approach.

The frequentist procedures of significance testing and confidence intervals can be constructed without regard to utility functions. However, some elements of frequentist statistics, such as statistical decision theory, do incorporate utility functions. In particular, frequentist developments of optimal inference (such as minimum-variance unbiased estimators, or uniformly most powerful testing) make use of loss functions, which play the role of (negative) utility functions. Loss functions need not be explicitly stated for statistical theorists to prove that a statistical procedure has an optimality property. However, loss-functions are often useful for stating optimality properties: for example, median-unbiased estimators are optimal under absolute value loss functions, in that they minimize expected loss, and least squares estimators are optimal under squared error loss functions, in that they minimize expected loss.

While statisticians using frequentist inference must choose for themselves the parameters of interest, and the estimators/test statistic to be used, the absence of obviously explicit utilities and prior distributions has helped frequentist procedures to become widely viewed as 'objective'.

Bayesian Inference

The Bayesian calculus describes degrees of belief using the 'language' of probability; beliefs are positive, integrate to one, and obey probability axioms. Bayesian inference uses the available posterior beliefs as the basis for making statistical propositions. There are several different justifications for using the Bayesian approach.

Examples of Bayesian Inference

- Credible intervals for interval estimation
- Bayes factors for model comparison

Bayesian Inference, Subjectivity and Decision Theory

Many informal Bayesian inferences are based on "intuitively reasonable" summaries of the posterior. For example, the posterior mean, median and mode, highest posterior density intervals, and Bayes Factors can all be motivated in this way. While a user's utility function need not be stated for this sort of inference, these summaries do all depend (to some extent) on stated prior beliefs, and are generally viewed as subjective conclusions. (Methods of prior construction which do not require external input have been proposed but not yet fully developed.)

Formally, Bayesian inference is calibrated with reference to an explicitly stated utility, or loss function; the 'Bayes rule' is the one which maximizes expected utility, averaged over the posterior uncertainty. Formal Bayesian inference therefore automatically provides optimal decisions in a decision theoretic sense. Given assumptions, data and utility, Bayesian inference can be made for essentially any problem, although not every statistical inference need have a Bayesian interpretation. Analyses which are not formally Bayesian can be (logically) incoherent; a feature of Bayesian procedures which use proper priors (i.e., those integrable to one) is that they are guaranteed to be coherent. Some advocates of Bayesian inference assert that inference *must* take place in this decision-theoretic framework, and that Bayesian inference should not conclude with the evaluation and summarization of posterior beliefs.

Other Modes of Inference (Besides Frequentist and Bayesian)

Information and Computational Complexity

Other forms of statistical inference have been developed from ideas in information theory and the theory of Kolmogorov complexity. For example, the minimum description length (MDL) principle selects statistical models that maximally compress the data; inference proceeds without assuming counterfactual or non-falsifiable 'data-generating mechanisms' or probability models for the data, as might be done in frequentist or Bayesian approaches.

However, if a 'data generating mechanism' does exist in reality, then according to Shannon's source coding theorem it provides the MDL description of the data, on average and asymptotically. In minimizing description length (or descriptive complexity), MDL estimation is similar to maximum likelihood estimation and maximum a posteriori estimation (using maximum-entropy Bayesian priors). However, MDL avoids assuming that the underlying probability model is known; the MDL principle can also be applied without assumptions that e.g. the data arose from independent sampling. The MDL principle has been applied in communication-coding theory in information theory, in linear regression, and in time-series analysis (particularly for choosing the degrees of the polynomials in Autoregressive moving average (ARMA) models). Information-theoretic statistical inference has been popular in data mining, which has become a common approach for very large observational and heterogeneous datasets made possible by the computer revolution and internet.

The evaluation of statistical inferential procedures often uses techniques or criteria from computational complexity theory or numerical analysis.

Fiducial Inference

Fiducial inference was an approach to statistical inference based on fiducial probability, also known as a "fiducial distribution". In subsequent work, this approach has been called ill-defined, extremely limited in applicability, and even fallacious. However this argument is the same as that which shows that a so-called confidence distribution is not a valid probability distribution and, since this has not invalidated the application of confidence intervals, it does not necessarily invalidate conclusions drawn from fiducial arguments.

Structural Inference

Developing ideas of Fisher and of Pitman from 1938 to 1939, George A. Barnard developed "structural inference" or "pivotal inference", an approach using invariant probabilities on group families. Barnard reformulated the arguments behind fiducial inference on a restricted class of models on which "fiducial" procedures would be well-defined and useful.

Statistical Assumptions

Statistical assumptions are general assumptions about statistical populations.

Statistics, like all mathematical disciplines, does not generate valid conclusions from nothing. Generating interesting conclusions about real statistical populations usually requires some background assumptions. These must be made carefully, because incorrect assumptions can generate wildly inaccurate conclusions.

The most commonly applied statistical assumptions are:

1. independence of observations from each other: This assumption is a common error.

2. independence of observational error from potential confounding effects
3. exact or approximate normality of observations: The assumption of normality is often erroneous, because many populations are not normal. However, it is standard practice to assume that the sample mean from a random sample is normal, because of the central-limit theorem.
4. linearity of graded responses to quantitative stimuli

Types of Assumptions

Statistical assumptions can be categorised into a number of types: Non-modelling assumptions. Statistical analyses of data involve making certain types of assumption, whether or not a formal statistical model is used. Such assumptions underlie even descriptive statistics.

Population assumptions. A statistical analysis of data is made on the basis that the observations available derive from either a single population or several different populations, each of which is, in some way, meaningful. Here a "population" is informally a set of other possible observations that might have been made. The assumption here is a simple one, to the effect that the observer should know that the observations obtained are representative of the problem, topic or class of objects being studied.

Sampling assumptions. These relate to the way observations have been gathered, and may often involve an assumption of random selection of some type. Modelling assumptions. These may be divided into three types: Distributional assumptions. Where a statistical model involves terms relating to random errors assumptions may be made about the probability distribution of these errors. In some cases, the distributional assumption relates to the observations themselves. Structural assumptions.

Statistical relationships between variables are often modelled by equating one variable to a function of another (or several others), plus a random error. Models often involve making a structural assumption about the form of the functional relationship here: for example, as in linear regression. This can be generalised to models involving relationships between underlying unobserved latent variables. Cross-variation assumptions. These assumptions involve the joint probability distributions of either the observations themselves or the random errors in a model. Simple models may include the assumption that observations or errors are statistically independent.

Checking Assumptions

Given that the validity of conclusions drawn from a statistical analysis depend on the validity of any assumptions made, it is clearly important that these assumptions should be reviewed at some stage. Some instances—for example where data are lacking—may require that researchers judge whether an assumption is reasonable. They can expanded this slightly to consider what effect a departure from the assumptions might produce. Where more extensive data are available, various types of procedure for statistical model validation are available, in particular for regression model validation.

Statistical Decision Theory

Decision theory in economics, psychology, philosophy, mathematics, and statistics is concerned with identifying the values, uncertainties and other issues relevant in a given decision, its rationality, and the resulting optimal decision. It is closely related to the field of game theory as to interactions of agents with at least partially conflicting interests whose decisions affect each other.

Normative and Descriptive Decision Theory

Most of decision theory is normative or prescriptive, *i.e.*, it is concerned with identifying the best decision to take (in practice, there are situations in which "best" is not necessarily the maximal, optimum may also include values in addition to maximum, but within a specific or approximative range), assuming an ideal decision maker who is fully informed, able to compute with perfect accuracy, and fully rational. The practical application of this prescriptive approach (how people *ought to* make decisions) is called decision analysis, and aimed at finding tools, methodologies and software to help people make better decisions. The most systematic and comprehensive software tools developed in this way are called decision support systems.

Since people usually do not behave in ways consistent with axiomatic rules, often their own, leading to violations of optimality, there is a related area of study, called a positive or descriptive discipline, attempting to describe what people will actually do. Since the normative, optimal decision often creates hypotheses for testing against actual behaviour, the two fields are closely linked. Furthermore it is possible to relax the assumptions of perfect information, rationality and so forth in various ways, and produce a series of different prescriptions or predictions about behaviour, allowing for further tests of the kind of decision-making that occurs in practice.

In recent decades, there has been increasing interest in what is sometimes called 'behavioural decision theory' and this has contributed to a re-evaluation of what rational decision-making requires.

What Kinds of Decisions Need a Theory?

Choice Under Uncertainty

This area represents the heart of decision theory. The procedure now referred to as expected value was known from the 17th century. Blaise Pascal invoked it in his famous wager, which is contained in his *Pensées*, published in 1670. The idea of expected value is that, when faced with a number of actions, each of which could give rise to more than one possible outcome with different probabilities, the rational procedure is to identify all possible outcomes, determine their values (positive or negative) and the probabilities that will result from each course of action, and multiply the two to give an expected value. The action to be chosen should be the one that gives rise to the highest total expected value. In 1738, Daniel Bernoulli published an influential paper entitled *Exposition of a New Theory on the Measurement of Risk*, in which he uses the St. Petersburg paradox to show that expected value theory must be normatively wrong. He also gives an example in which a Dutch merchant is trying to decide whether to insure a cargo being sent from Amsterdam to St Petersburg in winter, when it is known that there is a 5% chance that the ship and cargo will be lost. In his solution, he defines a utility function and computes expected utility rather than expected financial value.

In the 20th century, interest was reignited by Abraham Wald's 1939 paper pointing out that the two central procedures of sampling–distribution–based statistical-theory, namely hypothesis testing and parameter estimation, are special cases of the general decision problem. Wald's paper renewed and synthesized many concepts of statistical theory, including loss functions, risk functions, admissible decision rules, antecedent distributions, Bayesian procedures, and minimax procedures. The phrase "decision theory" itself was used in 1950 by E. L. Lehmann.

The revival of subjective probability theory, from the work of Frank Ramsey, Bruno de Finetti, Leonard Savage and others, extended the scope of expected utility theory to situations where subjective probabilities can be used. At this time, von Neumann's theory of expected utility proved that expected utility maximization followed from basic postulates about rational behaviour.

The work of Maurice Allais and Daniel Ellsberg showed that human behaviour has systematic and sometimes important departures from expected-utility maximization. The prospect theory of Daniel Kahneman and Amos Tversky renewed the empirical study of economic behaviour with less emphasis on rationality presuppositions. Kahneman and Tversky found three regularities — in actual human decision-making, "losses loom larger than gains"; persons focus more on *changes* in their utility–states than they focus on absolute utilities; and the estimation of subjective probabilities is severely biased by anchoring.

Castagnoli and LiCalzi (1996), Bordley and LiCalzi (2000) recently showed that maximizing expected utility is mathematically equivalent to maximizing the probability that the uncertain consequences of a decision are preferable to an uncertain benchmark (e.g., the probability that a mutual fund strategy outperforms the S&P 500 or that a firm outperforms the uncertain future performance of a major competitor.). This reinterpretation relates to psychological work suggesting that individuals have fuzzy aspiration levels (Lopes & Oden), which may vary from choice context to choice context. Hence it shifts the focus from utility to the individual's uncertain reference point. Pascal's Wager is a classic example of a choice under uncertainty.

Intertemporal Choice

This area is concerned with the kind of choice where different actions lead to outcomes that are realised at different points in time. If someone received a windfall of several thousand dollars, they could spend it on an expensive holiday, giving them immediate pleasure, or they could invest it in a pension scheme, giving them an income at some time in the future. What is the optimal thing to do? The answer depends partly on factors such as the expected rates of interest and inflation, the person's life expectancy, and their confidence in the pensions industry. However even with all those factors taken into account, human behaviour again deviates greatly from the predictions of prescriptive decision theory, leading to alternative models in which, for example, objective interest rates are replaced by subjective discount rates.

Competing Decision Makers

Some decisions are difficult because of the need to take into account how other people in the situation will respond to the decision that is taken. The analysis of such social decisions is more often treated under the label of game theory, rather than decision theory, though it involves the same mathematical methods. From the standpoint of game theory most of the problems treated in decision theory are one-player games (or the one player is viewed as playing against an impersonal background situation). In the emerging socio-cognitive engineering, the research is especially focused on the different types of distributed decision-making in human organisations, in normal and abnormal/ emergency/crisis situations. Signal detection theory is based on decision theory.

Complex Decisions

Other areas of decision theory are concerned with decisions that are difficult simply because of their complexity, or the complexity of the organisation that has to make them. In such cases the issue is not the deviation between real and optimal behaviour, but the difficulty of determining the optimal behaviour in the first place. The Club of Rome, for example, developed a model of economic growth and resource usage that helps politicians make real-life decisions in complex situations. Decisions are also affected by whether options are framed together or separately. This is known as the distinction bias.

The Compromise Effect

One area of decision deals with choosing between options involving extremes. When faced with choices, the decider will tend to favour options holding moderate benefits as opposed to options resulting in extreme consequences. The Compromise Effect operates under a mindset driven by the belief that the most moderate option, amid extremes, carries the most benefits from each extreme.

Heuristics

One method of decision-making is heuristic. The heuristic approach makes decisions based on routine thinking. While this is quicker than step-by-step processing, heuristic decision-making opens the risk of inaccuracy. Mistakes that otherwise would have been avoided in step-by-step processing can be made. One common and incorrect thought process that results from heuristic thinking is the gambler's fallacy. The gambler's fallacy makes the mistake of believing that a random event is affected by previous random events. For example, there is a fifty percent chance of a coin landing on heads. Gambler's fallacy suggests that if the coin lands on tails, the next time it flips, it will land on heads, as if it's "the coin's turn" to land on heads. This is simply not true. Such a fallacy is easily disproved in a step-by-step process of thinking.

Alternatives to Decision Theory

A highly controversial issue is whether one can replace the use of probability in decision theory by other alternatives.

Probability Theory

The Advocates of probability theory point to: the work of Richard Threlkeld Cox for justification of the probability axioms, the Dutch book paradoxes of Bruno de Finetti as illustrative of the theoretical difficulties that can arise from departures from the probability axioms, and the complete class theorems, which show that all admissible decision rules are equivalent to the Bayesian decision rule for some utility function and some prior distribution (or for the limit of a sequence of prior distributions). Thus, for every decision rule, either the rule may be reformulated as a Bayesian procedure (or a limit of a sequence of such), or there is a rule that is sometimes better and never worse.

Alternatives to Probability Theory

The proponents of fuzzy logic, possibility theory, quantum cognition, Dempster–Shafer theory, and info-gap decision theory maintain that probability is only one of many alternatives and point to many examples where non-standard alternatives have been implemented with apparent success; notably, probabilistic decision theory is sensitive to assumptions about the probabilities of various events, while non-probabilistic rules such as minimax are robust, in that they do not make such assumptions.

General Criticism

A general criticism of decision theory based on a fixed universe of possibilities is that it considers the "known unknowns", not the "unknown unknowns": it focuses on expected variations, not on unforeseen events, which some argue (as in black swan theory) have outsized impact and must be considered – significant events may be "outside model". This line of argument, called the ludic fallacy, is that there are inevitable imperfections in modelling the real world by particular models, and that unquestioning reliance on models blinds one to their limits.

Estimation Theory

Estimation theory is a branch of statistics that deals with estimating the values of parameters based on measured/empirical data that has a random component. The parameters describe an underlying physical setting in such a way that their value affects the distribution of the measured data. An estimator attempts to approximate the unknown parameters using the measurements. For example, it is desired to estimate the proportion of a population of voters who will vote for a particular candidate. That proportion is the parameter sought; the estimate is based on a small random sample of voters.

Or, for example, in radar the goal is to estimate the range of objects (airplanes, boats, etc.) by analyzing the two-way transit timing of received echoes of transmitted pulses. Since the reflected pulses are unavoidably embedded in electrical noise, their measured values are randomly distributed, so that the transit time must be estimated.

In estimation theory, two approaches are generally considered.

- The probabilistic approach assumes that the measured data is random with probability distribution dependent on the parameters of interest
- The set-membership approach assumes that the measured data vector belongs to a set which depends on the parameter vector.

For example, in electrical communication theory, the measurements which contain information regarding the parameters of interest are often associated with a noisy signal. Without randomness, or noise, the problem would be deterministic and estimation would not be needed.

Basics

To build a model, several statistical "ingredients" need to be known. These are needed to ensure the estimator has some mathematical tractability.

The first is a set of statistical samples taken from a random vector (RV) of size N. Put into a vector,

$$\mathbf{x} = \begin{bmatrix} x[0] \\ x[1] \\ \vdots \\ x[N-1] \end{bmatrix}.$$

Secondly, there are the corresponding M parameters

$$\boldsymbol{\theta} = \begin{bmatrix} \theta_1 \\ \theta_2 \\ \vdots \\ \theta_M \end{bmatrix},$$

which need to be established with their continuous probability density function (pdf) or its discrete counterpart, the probability mass function (pmf)

$$p(\mathbf{x} \mid \boldsymbol{\theta}).$$

It is also possible for the parameters themselves to have a probability distribution (e.g., Bayesian statistics). It is then necessary to define the Bayesian probability

$$\pi(\boldsymbol{\theta}).$$

After the model is formed, the goal is to estimate the parameters, commonly denoted $\hat{\boldsymbol{\theta}}$, where the "hat" indicates the estimate.

One common estimator is the minimum mean squared error estimator, which utilizes the error between the estimated parameters and the actual value of the parameters

$$\mathbf{e} = \hat{\boldsymbol{\theta}} - \boldsymbol{\theta}$$

as the basis for optimality. This error term is then squared and minimized for the MMSE estimator.

Maximum of a Uniform Distribution

One of the simplest non-trivial examples of estimation is the estimation of the maximum of a uniform distribution. It is used as a hands-on classroom exercise and to illustrate basic principles of estimation theory. Further, in the case of estimation based on a single sample, it demonstrates philosophical issues and possible misunderstandings in the use of maximum likelihood estimators and likelihood functions.

Given a discrete uniform distribution $1, 2, \ldots, N$ with unknown maximum, the UMVU estimator for the maximum is given by

$$\frac{k+1}{k} m - 1 = m + \frac{m}{k} - 1$$

where m is the sample maximum and k is the sample size, sampling without replacement. This problem is commonly known as the German tank problem, due to application of maximum estimation to estimates of German tank production during World War II.

The formula may be understood intuitively as:

"The sample maximum plus the average gap between observations in the sample", the gap being added to compensate for the negative bias of the sample maximum as an estimator for the population maximum.

This has a variance of

$$\frac{1}{k}\frac{(N-k)(N+1)}{(k+2)} \approx \frac{N^2}{k^2} \text{ for small samples } k \ll N$$

so a standard deviation of approximately N/k, the (population) average size of a gap between samples; compare $\frac{m}{k}$ above. This can be seen as a very simple case of maximum spacing estimation.

The sample maximum is the maximum likelihood estimator for the population maximum, but, as discussed above, it is biased.

Applications

Numerous fields require the use of estimation theory. Some of these fields include (but are by no means limited to):

- Interpretation of scientific experiments
- Signal processing
- Clinical trials
- Opinion polls
- Quality control
- Telecommunications
- Project management
- Software engineering
- Control theory (in particular Adaptive control)
- Network intrusion detection system
- Orbit determination

Measured data are likely to be subject to noise or uncertainty and it is through statistical probability that optimal solutions are sought to extract as much information from the data as possible.

Statistical Hypothesis Testing

A statistical hypothesis concerns the behaviour of observable random variables; i.e. it is a scientific hypothesis that is testable on the basis of observing a process that is modelled via a set of random variables. A statistical hypothesis test is a method of statistical inference used for testing a statistical hypothesis.

A result is called *statistically significant* if it has been predicted as unlikely to have occurred by chance alone, according to a predetermined threshold probability, the significance level. Hypothesis tests are used in determining what outcomes of a study would lead to a rejection of the null hypothesis for a pre-specified level of significance; this can help to decide whether results contain enough information to reject the null hypothesis. In the Neyman-Pearson framework, the process of distinguishing between the null hypothesis and the alternative hypothesis is aided by identifying two conceptual types of errors (type 1 & type 2), and by specifying parametric limits on e.g. how much type 1 error will be permitted.

An alternative framework for statistical hypothesis testing is to specify a set of statistical models, one for each candidate hypothesis, and then use statistical model selection techniques to choose the most appropriate model. The most common selection techniques are based on either Akaike information criterion or Bayes factor.

Statistical hypothesis testing is sometimes called confirmatory data analysis. It can be contrasted with exploratory data analysis, which may not have pre-specified hypotheses.

Variations and Sub-classes

Statistical hypothesis testing is a key technique of both Frequentist inference and Bayesian inference although they have notable differences. Statistical hypothesis tests define a procedure that controls (fixes) the probability of incorrectly *deciding* that a default position (null hypothesis) is incorrect based on how likely it would be for a set of observations to occur if the null hypothesis were true. Note that this probability of making an incorrect decision is not the probability that the null hypothesis is true, nor whether any specific alternative hypothesis is true. This contrasts with other possible techniques of decision theory in which the null and alternative hypothesis are treated on a more equal basis. One naive Bayesian approach to hypothesis testing is to base decisions on the posterior probability, but this fails when comparing point and continuous hypotheses. Other approaches to decision making, such as Bayesian decision theory, attempt to balance the consequences of incorrect decisions across all possibilities, rather than concentrating on a single null hypothesis. A number of other approaches to reaching a decision based on data are available via decision theory and optimal decisions, some of which have desirable properties, yet hypothesis testing is a dominant approach to data analysis in many fields of science. Extensions to the theory of hypothesis testing include the study of the power of tests, the probability of correctly rejecting the null hypothesis given that it is false. Such considerations can be used for the purpose of sample size determination prior to the collection of data.

The Testing Process

In the statistics literature, statistical hypothesis testing plays a fundamental role. The usual line of reasoning is as follows:

1. There is an initial research hypothesis of which the truth is unknown.
2. The first step is to state the relevant null and alternative hypotheses. This is important as misstating the hypotheses will muddy the rest of the process.
3. The second step is to consider the statistical assumptions being made about the sample in doing the test; for example, assumptions about the statistical independence or about the form of the distributions of the observations. This is equally important as invalid assumptions will mean that the results of the test are invalid.
4. Decide which test is appropriate, and state the relevant test statistic T.
5. Derive the distribution of the test statistic under the null hypothesis from the assumptions. In standard cases this will be a well-known result. For example the test statistic might follow a Student's t distribution or a normal distribution.
6. Select a significance level (α), a probability threshold below which the null hypothesis will be rejected. Common values are 5% and 1%.
7. The distribution of the test statistic under the null hypothesis partitions the possible values of T into those for which the null hypothesis is rejected—the so-called *critical region*—and those for which it is not. The probability of the critical region is α.
8. Compute from the observations the observed value t_{obs} of the test statistic T.
9. Decide to either reject the null hypothesis in favour of the alternative or not reject it. The decision rule is to reject the null hypothesis H_0 if the observed value t_{obs} is in the critical region, and to accept or "fail to reject" the hypothesis otherwise.

An Alternative Process is Commonly Used

1. Compute from the observations the observed value t_{obs} of the test statistic T.
2. Calculate the p-value. This is the probability, under the null hypothesis, of sampling a test statistic at least as extreme as that which was observed.

3. Reject the null hypothesis, in favour of the alternative hypothesis, if and only if the p-value is less than the significance level (the selected probability) threshold.

The two processes are equivalent. The former process was advantageous in the past when only tables of test statistics at common probability thresholds were available. It allowed a decision to be made without the calculation of a probability. It was adequate for classwork and for operational use, but it was deficient for reporting results.

The latter process relied on extensive tables or on computational support not always available. The explicit calculation of a probability is useful for reporting. The calculations are now trivially performed with appropriate software.

The difference in the two processes applied to the Radioactive suitcase example (below):

- "The Geiger-counter reading is 10. The limit is 9. Check the suitcase."
- "The Geiger-counter reading is high; 97% of safe suitcases have lower readings. The limit is 95%. Check the suitcase."

The former report is adequate, the latter gives a more detailed explanation of the data and the reason why the suitcase is being checked.

It is important to note the philosophical difference between accepting the null hypothesis and simply failing to reject it. The "fail to reject" terminology highlights the fact that the null hypothesis is assumed to be true from the start of the test; if there is a lack of evidence against it, it simply continues to be assumed true. The phrase "accept the null hypothesis" may suggest it has been proved simply because it has not been disproved, a logical fallacy known as the argument from ignorance. Unless a test with particularly high power is used, the idea of "accepting" the null hypothesis may be dangerous. Nonetheless the terminology is prevalent throughout statistics, where its meaning is well understood.

The processes described here are perfectly adequate for computation. They seriously neglect the design of experiments considerations.

It is particularly critical that appropriate sample sizes be estimated before conducting the experiment.

The phrase "test of significance" was coined by statistician Ronald Fisher.

Interpretation

If the p-value is less than the required significance level (equivalently, if the observed test statistic is in the critical region), then we say the null hypothesis is rejected at the given level of significance. Rejection of the null hypothesis is a conclusion. This is like a "guilty" verdict in a criminal trial – the evidence is sufficient to reject innocence, thus proving guilt. We might accept the alternative hypothesis (and the research hypothesis).

If the p-value is not less than the required significance level (equivalently, if the observed test statistic is outside the critical region), then the test has no result. The evidence is insufficient to support a conclusion. (This is like a jury that fails to reach a verdict.) The researcher typically gives extra consideration to those cases where the p-value is close to the significance level.

In the Lady tasting tea example (below), Fisher required the Lady to properly categorize all of the cups of tea to justify the conclusion that the result was unlikely to result from chance. He defined the critical region as that case alone. The region was defined by a probability (that the null hypothesis was correct) of less than 5%.

Whether rejection of the null hypothesis truly justifies acceptance of the research hypothesis depends on the structure of the hypotheses. Rejecting the hypothesis that a large paw print originated from a bear does not immediately prove the existence of Bigfoot. Hypothesis testing emphasizes the rejection which is based on a probability rather than the acceptance which requires extra steps of logic.

"The probability of rejecting the null hypothesis is a function of five factors: whether the test is one- or two tailed, the level of significance, the standard deviation, the amount of deviation from the null hypothesis, and the number of observations." These factors are a source of criticism; factors under the control of the experimenter/analyst give the results an appearance of subjectivity.

Use and Importance

Statistics are helpful in analyzing most collections of data. This is equally true of hypothesis testing which can justify conclusions even when no scientific theory exists. In the Lady tasting tea example, it was "obvious" that no difference existed between (milk poured into tea) and (tea poured into milk). The data contradicted the "obvious".

Real world applications of hypothesis testing include:

- Testing whether more men than women suffer from nightmares
- Establishing authorship of documents

- Evaluating the effect of the full moon on behaviour
- Determining the range at which a bat can detect an insect by echo
- Deciding whether hospital carpeting results in more infections
- Selecting the best means to stop smoking
- Checking whether bumper stickers reflect car owner behaviour
- Testing the claims of handwriting analysts

Statistical hypothesis testing plays an important role in the whole of statistics and in statistical inference. Significance testing has been the favoured statistical tool in some experimental social sciences (over 90% of articles in the Journal of Applied Psychology during the early 1990s). Other fields have favoured the estimation of parameters (e.g., effect size). Significance testing is used as a substitute for the traditional comparison of predicted value and experimental result at the core of the scientific method. When theory is only capable of predicting the sign of a relationship, a directional (one-sided) hypothesis test can be configured so that only a statistically significant result supports theory. This form of theory appraisal is the most heavily criticized application of hypothesis testing.

Chapter 8

Spatial Modelling and the Evolution of Spatial Theory

There are many definitions for the models varying as the following quote from Haggett and Chorley (1967) illustrates. A model may be defined as: „...either a theory, a law, a hypothesis or a structured idea. It can be a role, a relation or an equation. It can be a synthesis of data. Most important, from the geographical viewpoint, it can also include reasoning about the real world by means of translations in space (to give spatial models) or in time (to give historical models)" (Haggett and Chorley 1967, p. 21-22.)

As Haines-Young and Petch (1986) indicates the idea that a model is an idealized or simplified representation of reality is widely held. Every approaches attempt to simplify reality by abstracting only those parts of the real world which were thought to be essential in defining and solving the problem. In addition, a set of structured ideas has been used to develop a solution to the problem being addressed.

Spatial modelling, as the definition by Haggett and Chorley (1967) illustrates, is simply one component of the modelling process. In this process, geographical data about the spatial relationship of features in the real world are used to help us understand and address a particular problem. This might be as simple as how to get from A to B by the shortest route or as complex as a model of the atmosphere designed to simulate global warming.

As Haines-Young and Petch state (1986) a model maybe as complex or as simplistic as the user requires. However, in all cases a model is an abstraction of the real world, which provides a framework for understanding real world processes and, perhaps more importantly, planning how to manage them. The building of models is a continuous iterative process, because as we gain knowledge about the real world from applying the model, we are able to identify the weaknesses in our model and change it. Understanding the dynamic nature of model building is important because it reminds us that using a GIS is a process, which will be open to change and review. As you revisit the way your organisation handles and uses spatial information in a GIS so you will wish to adapt the model upon which your GIS has been constructed.

Building Spatial Model

Before dealing in detail the spatial model building, let's look at three types of modelling, which are known less or more from our earlier studies: - The map - The architectural model - The aerial photographs

The Map as a Model

The map has used as a model for the spatial data storage and display, since the human can draw marks on a surface. The first maps drawn on the cave walls incised with sticks or charcoals. From our perspective the question is, how to use the map to display elements the real world in two-dimension space. In this figure, the map is represents, how to get to a spectacle in the town.

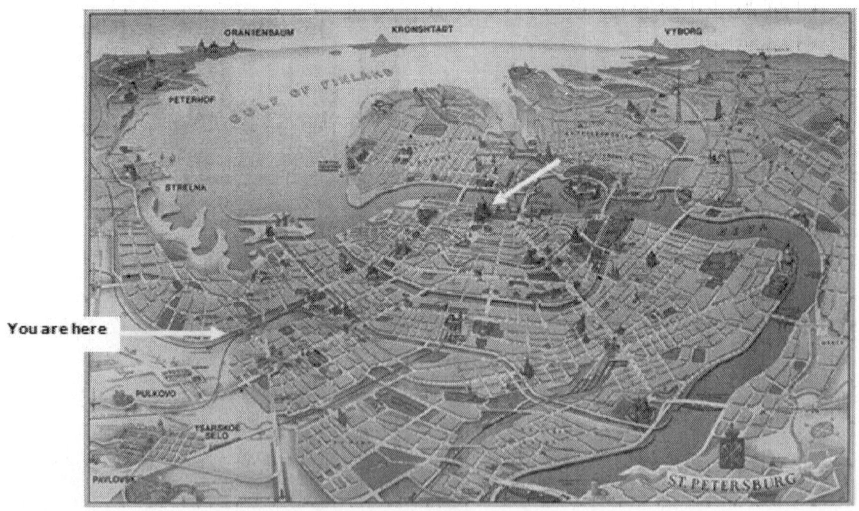

Figure: *The cartographic model This figure shows an example, how to use the map as a model of reality. The sketch give an idea, how can be to get to the cathedral (arrow) from our starting point (arrow). We can imagine in advance what kind of objects we will see along the way (typical buildings), when and where to turn to the road and what buildings and landmarks will be compared to find the cathedral. If you are just a little familiar with the area and the interpretation of symbols, you will find there. This*

drawing is simplifies and extracts reality. In other words, the author has created a very simple but very effective model for the real world. As with all models, the efficiency here based on the user's assumed knowledge. In the above example, the author assumes that we can read text, understand symbols, and can orient in space on the basis of the detected objects.

Assumed knowledge and the proper use of two-dimensional models is an issue that playing central role in the practical use of GIS.

For better understanding how the map works like a two-dimensional model of reality, let us observe a topographic map.

Figure: *Part of topographic map (onp.nemzetipark.gov.hu)*

The map uses symbols, shapes, lines, colours and text for display the elements of the real world. This map is including only „necessary" individuals, such as roads, railways, houses, rivers, flora and fauna. Every element of the map is only approach of the real features. We will do not find a tree, where on the map is drawn a wooden sign. Outline of the forest is much more complicated in reality, as seen on the map. The roads are in the correct place, but their width is exaggerated. If you measure the width of the road on the map, you get a much larger number than is realistic. Using the map assumes that we can understand how it might look like the reality by our experience. Each map's purpose is to convey information. The type of conveyed information is changing map by map. A topographic map is an excellent example of the orientation, hiking, to organise transport etc.

The Architectural Model

The architectural model or any other model, which uses a third dimension to visualize the real world, is probably the easiest way to understand the spatial model. The reason is that this model is closer to the human form of approach, instead of the two-dimensional map.

The map and the architectural model convey of the real world like lines, symbols and shapes. Aerial photographs however are up different shadows, colours and shades. One of the characteristic observation is that the boundaries between features are uncertain on aerial photographs, and consist colours, shades etc. The borders on architectural model or on map are defined more precisely. Think of a sharp corner of a building or plot of land with a straight boundary line. One can wonder whether the sharp demarcation or insecure is closer to reality. The aerial photographs are a source of data used in GIS and its correct interpretation is a specific profession.

Modelling the Spatial Dimension

All three of the spatial models we have looked at above; the air photograph, the architects' model and the map, address two dimensions of reality:

The Spatial Dimension; Where Something is Located

The thematic dimension; the character of either the location or the object occupying that location and they could (if we had a sequence of them) have been used to address a third:

The Temporal Dimension; the Comparison of Data over Time

In the rest of this module, we will concentrate on how GIS systems model the spatial dimension. Central to the concept of developing spatial models is the notion of capturing spatial data. Usually this is done through a process of observation and measurement. Here a problem arises because as Burrough (1987) says,

„Unlike many other kinds of information handled routinely by modern information systems (computers), geographical data are complicated by the fact that they must include information about position, possible topological connections, and attributes of the objects recorded."

Let us examine again briefly, what we mean by the term spatial data. It is useful to start with a definition of data itself. Everest (1986) provides a concise and useful definition:

> *"Data are 'facts' represented by values, numbers and character strings or symbols which carry meaning in a certain context."*

We can therefore consider spatial data as values, character strings or symbols, which convey to their user information about the geographical location of entities that can be observed in the real world. An additional element of spatial data is that they must have a pointer that describes their location on the earth's surface. Probably one of the most commonly used pointers for spatial data is an address. This alphanumeric reference is used as a code to describe the location of an individual premise. Another common pointer used is the coordinate reference. Both these pointers represent spatial referencing systems. The spatial referencing of geographical data is an essential component of GIS.

As Everest's definition suggests, data need not necessarily be number or characters. They could be symbols. Here a set of symbols has been used in conjunction with numbers and character strings to convey location information about the location of real features.

The data and information must be distinguished from each other clearly. The data is the form of encoded information, while deducing information from the data is used in solving the problem.

Building Spatial Models for the Computer Environment

The starting point for all GIS projects is what we call data model. If anyone is familiar with the database design, familiar with the concept of data model. The data model is basically a structure that the computer uses to simulate the problems. The GIS data modelling is the process, when the computer trying to solve a geographical problem. The data model can also be defined as a specific group of individuals and the general description of the relationships between individuals.

By geographical concepts the feature could be, for example, house, land or river. An important feature of the specimen is to clearly identify and distinguish it from other feature. The spatial elements of a relationship between entities in a geographical data model would be things such as proximity, adjacency, containment and direction. The important thing about entities and relationship is that they can both have attributes associated with them; for example the size of a house, how many windows it has, the number of doors, it's colour, value, type of building material. The attributes of a relationship may be the units in which it is expressed. Distance, for example, may have a value in time, or metres.

As Peuquet (1990) notes many data model designers realise that data needs to be viewed at a number of levels that progress from reality, through the abstract, user oriented information structure, to the concrete, machine-oriented storage structure of the computer. The data modelling process therefore can be summarized as a series of stages in data abstraction. With table, we move further from the real world closer to the computer representation of the geographical features we wish to model.

Chapter 9

Challenges in Spatial Data Analysis

This is a time of unprecedented opportunity for spatial analysis. More people than ever have access to the Global Positioning System for direct measurement of location on the

Earth's surface; to the products of high-resolution remote-sensing satellites; and to the manipulative power of geographic information systems (GIS). Some of these technologies are encountered in everyday life, through sites such as Google Earth, Google Maps, and Microsoft Windows Live Local, and through the widespread use of in -vehicle navigation systems.

Several academic disciplines are recording a spatial turn, a new and in some cases renewed interest in space and location as a framework for analysis, understanding, and presentation of results.

A recent publication (National Research Council, 2006) has defined and explored spatial thinking as a paradigm for primary and secondary education, and projects have been funded around the world to advance spatial literacy.

At the same time the field faces substantial challenges, as it attempts to take advantage of these new opportunities.

Computing and Networking Technology

In the early 1990s a substantial literature accumulated on the opportunities offered by GIS. In 1988 the U.S. National Science Foundation had established the National Centre for Geographic Information and Analysis (NCGIA) at three sites: the University of

California, Santa Barbara; the State University of New York at Buffalo; and the University of Maine. One of NCGIA's objectives was to advance the use of GIS across the sciences, as a platform for spatial analysis, so it was considered important to assess progress to date, and to identify and remove impediments to the greater use of spatial analysis. NCGIA organised a specialist meeting on the topic that eventually led to a book, and several additional papers appeared Underlying this spate of funding and writing was the simple premise that GIS provided an ideal means of implementing the known techniques of spatial analysis, as well as techniques that might be developed in the future. A single package, if sufficiently sophisticated, could offer easy and largely painless access to an abundance of robust, scientifically sound techniques for analyzing and visualizing spatial data.

The results of each stage of analysis could be fed into further stages, and data could be managed within a single environment that recognised a range of data formats.

Comparisons were frequently drawn with the statistical packages, which similarly offered easy access to a multitude of statistical techniques, along with the necessary housekeeping functions.

At the time, each GIS software product was organised into a single, monolithic package.

In the 1980s such packages were typically installed on minicomputers such as the VAX or Prime, but in the late 1980s the transition to personal Unix workstations and later to the PC and Mac had opened the possibility of an entirely individualized toolbox installed on the researcher's desk. GIS was likened to a butler –an intelligent assistant working with the user to solve problems, knowing the foibles and preferences of the user, and taking on those tasks that the user found too complex, tedious, time-consuming, or inaccurate if performed by hand. Abler hailed GIS as geography's equivalent of the microscope or the telescope, a powerful tool that allowed researchers to gain insights that were simply impossible with the normal senses and intuition.

From this perspective, the power of GIS would be judged simply by the proportion of known techniques of spatial analysis that it supported, by the accuracy with which it implemented each method, and by the extent to which it prevented misuse and misinterpretation of results. There were many complaints about this time regarding the success of GIS against these objectives. Commercial software developers were seen as insufficiently interested in supporting advanced spatial analysis, being content instead to direct their efforts at satisfying the needs of their more wealthy corporate and agency customers, whose interests tended to be more in data management and inventory. GIS designers failed to ground their products in sound theory, preferring intuitive terms and explanations over formal and mathematical ones. Because of this lack of formal grounding, each vendor tended to adopt its own terms, formats, and structures, leading to endless proliferation and an apparently insurmountable lack of interoperability.

It was in this context that the Web appeared on the scene, and the Internet emerged as the dominant and indeed quickly the only network for computer communication. Since 1993 and the release of Mosaic the impact of communications technology has been so profound as to change the entire landscape of GIS and spatial analysis. Sui and Goodchild have argued that the metaphor of the butler is no longer appropriate – instead, GIS technology now constitutes a medium through which people communicate what they know about the Earth's surface that is comparable to traditional media such as print, radio, and television. As such, its issues are dramatically different from those of earlier decades. Bandwidth, interoperability, and metadata have largely replaced computing speed, storage capacity, and the sophistication of desktop software as major concerns of GIS users. Even the most sophisticated of users no longer program, relying instead on the incredibly abundant resources of the Web, easy mechanisms for sharing code, and new forms of software architecture.

The following three sections explore some of these issues, and their implications for spatial analysis.

Server GIS

In the client-server computing paradigm that underlies the Web, the user or client's hardware and software are comparatively simple or thin, and most actual computation occurs remotely on a more powerful server. In the extreme, the user needs only a Web browser such as Microsoft Explorer or Netscape. Instead of installing a thick piece of software, such as a GIS package, the user obtains many if not all of its services from a remote server. For example, the task of finding the optimum route from an origin to a destination through a street network, the task per formed by many Web sites such as mapquest.com, no longer requires the user to obtain a powerful GIS and the necessary database representing roads and streets, and to mount both on his or her desktop machine, since the same service can be obtained free from the server. The user need only specify the origin and destination to the server using a Web browser; the results are then sent back from the server and displayed locally using the same Web browser. In principle all GIS functions and all types of spatial analysis could be organised in this way.

Instead of installing and operating their own software, researchers could send data to sites where sophisticated forms of spatial analysis were performed. Researchers developing new forms of spatial analysis would find it far easier to offer their techniques as Web services than to engage in the time-consuming distribution of software, and users would benefit by not having to spend time obtaining, installing, and maintaining their own copies.

Server GIS is now common among public agencies interested in providing public access to their spatial data, along with simple capabilities for query and visual display. Many local governments provide access to their land-ownership and property taxation databases in this way, allowing users to query details of their own and other properties, using a map interface.

In practice, however, server GIS has had a limited impact to date, particularly for more sophisticated analysis, for a number of reasons:

- There is no consensus on the appropriate business model for server GIS. Desktop GIS software generates income for its developers through sales and licensing, providing a healthy income stream, and developers of new methods of spatial analysis have sometimes used this same approach. Users of server GIS typically expect services to be provided free, leaving the providers of such services to generate income through advertising and the licensing of services to third parties.

Routing services, for example, can be found embedded in the Web sites of on-line travel agencies and real-estate companies, presumably at some cost to them.

Moreover, the software for server GIS tends to be more expensive per copy than conventional desktop GIS software.

- Server GIS is most effective when the volume of data that needs to be input by the user is limited, and when the data needed are common to a large number of users and applications

. A routing service, for example, requires only an origin and destination, and uses a generic database of streets and roads stored at the server.

Moreover, such databases change frequently, and there are enormous economies of scale if all users can rely on a single version. Geo-coding or address matching, the task of converting street addresses to coordinates, has become a popular function for server GIS for the same reason.

- Lack of interoperability continues to be an issue for server GIS. There are no standards for the description of services, though several geo-portals now provide limited directories. Extensive reformatting may be needed to make data readable by remote services, and the results returned may similarly need to be reformatted to be useful locally.

The choice between local and server-based computing is a complex one, and developers and implementers of spatial-analytic routines will need to consider the options carefully. However, it is clear that the nature of computing is changing, as many services move to a central, server-based model.

Process Scripts

Research tends to proceed in stages, as problems are formulated, data are collected and checked, analysis is performed, and results are scrutinized. Each stage feeds forward to the next, and also back to the previous stages, as projects are rethought and as hypotheses are tested and modified. By the time the project is finally completed, the investigator may well have lost track of some of the stages, and may find it difficult to provide the necessary details in publications and reports. Somewhat paradoxically, the research community has invested heavily in the infrastructure to create and share data, and in the software to process them, but has not made similar investments in the techniques for management of the research process. The problem grows more severe as research becomes more collaborative, with many participants who may or may not communicate in person, and as the tools of research become more complex.

Against this background it is not surprising that many vendors of GIS and spatial-analytic packages have created macro-or scripting languages that allow researchers to express complex analyses as sequences of operations, and to store, manage, and execute such sequences as simple commands. A script in digital form is immediately more easily shared, managed, and documented than its equivalent in the jotted and invariably incomplete hand-notes of the researcher. Modern scripting languages allow complex hierarchical structures, since a single line in a script can invoke other scripts and programs, and allow sequences of operations to be repeated many times in such applications as Monte Carlo simulation.

However, the design of an appropriate scripting language is a very sophisticated task, requiring a high level of knowledge of the needs of the research community, across many disciplines and domains. Simple scripting languages merely allow the user to invoke any of the commands of the package, but more sophisticated languages imply a recognition of the fundamental elements from which complex spatial analyses are built. If the granularity of the scripting language is too coarse, researchers will find it too difficult to express the full range of applications –and if it is too detailed, the script will be unnecessarily long.

The work of Tomlin provided the first successful effort at a generic scripting language for GIS, albeit only for congruent layers of raster data. The language was adopted by several packages, and several extensions were made. Van Deursen analyzed the operations required to support dynamic modelling in a raster environment, including the implementation of finite-difference models, in what became the scripting language for PCRaster, a raster-based package heavily oriented towards environmental modelling. Takeyama and Couclelis described a sophisticated language for the manipulation of pairs of raster cells, providing support for the analysis of spatial interactions. More broadly, all of these approaches are strongly related to the languages developed in image processing, or image algebras.

To date, however, there have been no comparably ambitious efforts to devise languages for vector data, or for the broader framework that spans both discrete objects and continuous fields.

Dynamic GIS that addresses both space and time also lacks comprehensive scripting languages. The effectiveness of future spatial analysis clearly depends on the community's ability to devise simple yet comprehensive languages that can be used to describe and share computational methods. In the past, mathematics provided an adequate language, and models were effectively shared using algebraic representation, through the pages of learned journals and books. But today's computational environments present a somewhat different problem, since the language of mathematics lies too far from actual implementation, and cannot readily be used to express the entire algorithmic basis of spatial analysis.

Interchangeable Software Components

Early computer software was comprised of programs, integrated pieces of software that performed well-defined functions. Early GIS developed in this context, and by the early 1990s a fully featured GIS such as ESRI's ARC/INFO included millions of lines of code, all designed to be compiled and executed together to provide a single, integrated computing environment.

This approach to software was both redundant, in the sense that large amounts of code might never be executed by a given user, whose interests might focus only on a small number of functions; and costly, in the sense that it was difficult for programmers to pull pieces of code out of one package to be reused in another. Even today, the average user of a package such as Microsoft Word will likely never have invoked many of the functions in this very large and complex package.

Several attempts to break out of this mold were made in the 1980s and 1990s. One of the more successful was the concept of a subroutine library, a collection of standard routines that could be

Called by programs, avoiding the need for repetitive reprogramming. Subroutine libraries became common in areas such as statistics, since they allowed comparatively sophisticated users to develop new programs quickly, relying on standard subroutines for many of the program's functions. The idea was difficult to implement for less sophisticated users, however, since it required each to possess a substantial knowledge of programming.

Contemporary approaches to software emphasize a rather different approach, in which sections of reusable code, or components, can be freely combined during the execution of a program. Standards have been developed by vendors such as Microsoft that allow compliant components to be freely linked and executed. Ungerer and Goodchild describe one such application, in which ESRI's ArcGIS and Microsoft's Excel have been combined to solve a standard problem in areal interpolation. Functions that are native to the GIS, such as polygon overlay, are obtained from ArcGIS, while operations on tables, such as matrix multiplication, are obtained from Excel. The entire analysis is invoked through commands written in Visual Basic, a form of scripting language, though other general scripting languages such as Python might also be used. Both packages are compliant with the Microsoft COM standard, allowing the components that form the building blocks of each to be freely combined and executed.

Approaches such as these are breaking down the barriers that previously existed between different types of software –in this case, ArcGIS and Excel –and allowing much more flexible forms of analysis. They invite an entirely new approach to software design, in which fundamental components with widespread application are combined to meet the needs of specific applications. They also call for answers to a fundamental question: what are the basic building blocks of spatial analytic software, and to what extent are the operations invoked by each form of analysis common to more than one form? Perhaps they will lead eventually to a new approach to teaching in spatial analysis, in which these fundamental building blocks are the elements of a course, rather than the analytic methods themselves.

Time and Dynamics

Many authors have commented on the generally static nature of GIS, and the difficulty of representing time and dynamic phenomena.

Most attribute this to the legacy of the paper map, which inevitably emphasizes those aspects of the Earth's surface that remain relatively static, over such dynamic phenomena as events, transactions, and flows.

Several comprehensive reviews have appeared, and much progress has been made in building spatial databases that include time.

This same emphasis on the static is evident in the toolkit of spatial analysis, with its focus on cross-sectional data. In part this is due to the difficulty of creating and acquiring longitudinal data: to the administrative difficulties that statistical agencies face in funding and maintaining data-collection programs through time; to the changing nature of the Earth's surface, and the impact that this has on data-collection procedures and the definitions of reporting zones; and to the changing nature of human society, and its notoriously short attention span. Efforts such as the National Historic GIS project have attempted to overcome these difficulties, building systems that allow users to construct longitudinal series from the census for example, but they remain comparatively few and far between.

While much progress has been made, the analysis of spatio-temporal data remains a

comparatively underexplored area, and a source of substantial challenges for the community. The next two subsections address two of these in greater detail.

Fundamental Laws

Much of the nature of GIS and many of the architectural choices that have been made over the past several decades are ultimately attributable to the nature of the data themselves –the ways in which spatial data are special. Anselin has identified two general characteristics, and Goodchild has discussed several more.

Spatial dependence describes the widely observed tendency for the variance of spatial data to increase with distance. To paraphrase Tobler, nearby things are more similar than distant things, a principle that has become known as the First Law of Geography.

All of the methods used to represent geographic phenomena in GIS are to some extent reliant on the validity of this principle. For example, there would be no value in representing topography with isolines if elevation did not vary smoothly, and there would be no value in aggregating areas into contiguous regions if the latter could not be designed with relatively low within -region variance.

Anselin's second principle is spatial heterogeneity, the tendency for the Earth's surface to exhibit spatial non-stationarity. All of the various techniques developed over the past two decades for local spatial analysis are based on this principle, since they attempt to summarize what is true locally, rather than what is true globally. The Geographically Weighted Regression of Fotheringham, Brunsdon, and Charlton falls into this category, as do the LISA technique of Anselin and the local statistics of Getis and Ord.

If such principles are generally true of spatial data, and are useful in guiding the development of computational systems, then one might reasonably ask whether similar principles exist for spatio-temporal data, and whether such principles might usefully inform the development of a more dynamic approach to GIS and spatial analysis. What is the spatio-temporal equivalent of Tobler's First Law, for example? Does spatial heterogeneity apply also in time? What relationships exist between the parameters of spatio-temporal and spatial dependence and heterogeneity? Are other general principles of spatio-temporal phenomena waiting to be discovered?

Dynamic Form

Spatial dependence and spatial heterogeneity are both properties of how the Earth's surface looks, capturing aspects of its form. Studies of form have a long history in science, but have given way in the long term to a desire to understand process –to understand how systems work, and the effects of human intervention. In geomorphology, for example, many scientists of the 19th and early 20th centuries were content to describe landforms, devising elaborate systems of morphological classification, and only later did interest develop in understanding how landforms came to be, and the processes that left such characteristic footprints on the surface. Today, of course, such studies of form are largely discredited, as they are in many other disciplines.

Because of its essentially static legacy, much GIS analysis has focused on form, and has been criticized for doing so. It is comparatively difficult to tease insights into process from cross- sectional form, though it is perhaps sometimes possible to eliminate false hypotheses about process. GIS has been accused of being the last manifestation of the quantitative revolution that occurred in geography in the 1960s, when Bunge and others attempted to draw insights from the similarity of forms found on the human and physical landscapes.

Very little is known, however, about the characteristic forms that may exist in spatio-temporal phenomena. Hagerstrand and others have examined the movements of individuals in space and time using three-dimensional displays, in which the two spatial dimensions form the horizontal plane and time forms the vertical axis. Much of this work focuses on similarities that may exist in the forms of such tracks, and the implications they may have for process. We know from the work of many researchers that different social conditions lead to dramatically distinct track forms, as for example in the differences between the daily tracks of single mothers, with their orientations to both workplace and daycare, and the tracks of workers in families in which only one of two adults works.

The development of greater support for time in GIS may lead to many other recognisable patterns in spatio-temporal data, and to a rebirth of interest in the study of spatio-temporal form. A new generation of analytic techniques is needed that extracts meaningful pattern from the mass of tracks displayed in the visualizations of Kwan and Lee and others, and links such patterns to hypotheses about process.

Spatial Literacy

In the past few years a remarkable series of Web sites have brought the sophisticated functions of GIS and spatial analysis much closer to the general public. While effective use of GIS requires extensive training, and in many cases advanced work at the undergraduate level, technologies such as Google Earth have given every citizen with a computer and a high-speed Internet connection access to many of the data sets and computational functions of GIS, and in some cases have even exposed the more sophisticated functions of spatial analysis. For example, anyone requesting driving directions from one of these sites receives answers that result from the execution of a complex algorithm that was previously the reserve of operations researchers and specialists in spatial optimization.

The methods of cartography and related disciplines are complex, and it is no surprise therefore that sophisticated tools in naïve hands can produce mistakes. A suitable example concerns the Greenwich meridian, and its position when displayed in Google Earth. Many users of this site have noted that the zero of longitude misses the Greenwich

Observatory by approximately 100m, and have posted comments, some of which conclude that a serious mistake has been made by Google, and by extension that the georegistration of imagery on the site is poor. In reality, the WGS84 (World Geodetic System of 1984) datum, now widely adopted around the world, does not place the Greenwich Observatory at exactly zero longitude, despite the international treaty that established it there in 1884 –and the position shown in Google Earth appears to be correct to within a few metres.

Although their support for spatial analysis is extremely limited, these sites have clearly provided the general public with access to a rich resource, and thousands of people have been empowered to create their own applications. The recent publication Mapping Hacks describes many fascinating examples, but contains not a single reference to the cartographic literature. At the same time students who have endured many hours of lectures and lab exercises to become competent in GIS may be frustrated to realise that a child of ten can create a computationally complex fly-by using Google Earth in a few minutes.

It seems clear that in part as a result of these developments the demand for basic knowledge of the principles of spatial analysis, GIS, geography, cartography, and related fields–for basic spatial literacy—is perhaps two or more orders of magnitude out of alignment with the supply. Education in these topics cannot be confined to a few advanced undergraduates, and to campuses lucky enough to have faculty interest, if it is to be accessible to the numbers of people now exposed to and enthusiastically adopting these tools. In this respect, spatial analysis faces an unprecedented challenge, to make itself known to a much larger community than previously.

There are several ways in which such a challenge might be met, by concerted effort on the part of the spatial-analysis community. One is to bring spatial literacy into the general-education or core curriculum of institutions of higher education, making its material accessible and eligible for credit for the vast majority of undergraduates.

Courses in other kinds of literacy are already available in this form; the argument needs to be made that familiarity with spatial analysis and GIS represents another, and arguably a more powerful form of literacy that should be part of the education of every citizen. Another strategy would be to develop a larger and more visible set of courses in the informal education sector, making spatial literacy part of on-line and certificate programs, and exposing its contents through libraries, museums, and other institutions. A third is to work to introduce spatial literacy earlier in the educational hierarchy, in high school and even elementary school. Valiant efforts have been made in this direction in the past, but they remain minimal in comparison with the size of the primary and secondary sectors, and there is much confusion about where such material might fit in the already stove-piped curriculum.

Beyond Traditional Practice in Science

When Harvey wrote his well-known and highly influential Explanation in Geography the dominant form of scientific practice centred on the individual investigator, whose methods followed a set of well-defined principles. For example, every experiment was to be reported in sufficient detail to allow its replication by another independent investigator. Every numerical result was to be reported with a level of precision that matched its accuracy. Every search of the literature was to be complete and comprehensive, so that the investigator could demonstrate knowledge of all previous and relevant work and prove the new work's originality. The principle of Occam's Razor –a willingness to adopt the simplest of several competing explanations –was universally accepted, as was the notion that all conclusions could be subject to empirical test and possible rejection. The goal of science was complete explanation, or in statistical terms an R2 of 1. When sample data were analyzed, all numerical results were to be subject to tests of statistical significance, to prove that they were not likely to be simply artifacts of the particular sample chosen, but properties of the population from which the sample was presumed to be drawn. All terms were to be rigorously defined, and vague terms were to be replaced by ones that met the standard of objectivity – rigorous and shared definition, such that two investigators would always agree on the outcome when the definition was applied.

These standards are of course collectively unattainable in all circumstances. They may be more attainable in some disciplines than others, and certainly it is possible to imagine a physicist having no difficulty adhering to them, and being fiercely critical of any study that appeared to relax them. But researchers in the general domain of this book clearly encounter situations in which one or more of them is distinctly problematic.

This is not to say that one should therefore reject them outright, and follow the lead of those who have looked for alternatives to scientific principles –rather, they constitute goals to which research should attempt always to aspire, while admitting that it may sometimes fall short.

Collaboration, Replicability and the Black Box

Before the widespread adoption of computing, it was customary for instructors in statistics courses to insist that each student be able to carry out a test by hand, before using any computational aids. Only then, it was argued, would the student fully understand the process involved, and be able to replicate it later. In this simple world it was possible to assume that every researcher knew every detail of every analysis, and that the published version of the research would include sufficient detail to allow others to repeat the experiment and replicate the results.

This principle has come under fire in recent decades, for a number of reasons. Computational aids have advanced to the point where it is not possible for any one individual to comprehend fully all of the algorithms involved. The author recalls passing a threshold, some time a round 1990, when it was no longer possible to believe that every aspect of a computational analysis could be replicated by hand, given enough time. Operating systems were perhaps the first such area of computing –by 1990 they had advanced to the point where it was no longer possible to believe they were the work of one person, or that any one individual fully understood every aspect of their operation.

Today these failures are commonplace. The documentation of our more sophisticated software, including GIS, is often not sufficient to detail every aspect of an analysis, and it

may be impossible to discover exactly how a given system computes a standard property, such as "slope", from a given input. In effect the developers of software, many of them operating in for-profit commercial environments, have become authorities that must be trusted, and it is difficult to submit their products to rigorous and exhaustive test.

Moreover, researchers now find it increasingly effective to work in teams, each team member providing some specific expertise. Funding agencies often express a willingness to fund research that brings together teams from many disciplines, in the interests of greater collaboration and cross-fertilization of ideas. But such arrangements inevitably lead to situations in which no one individual knows everything about an analysis, and members of the team have little alternative but to trust each other, just as researchers often have little alternative but to trust software.

Keeping the Stakeholders Happy

Tools such as GIS invite researchers to become involved in the processes of policy formulation and decision making.

The very architecture of GIS, with its database of local details and its procedures representing general principles, invites engagement with the ultimate users of research, since it allows decision makers to investigate the effects of manipulating outcomes in local contexts, and gives them many useful tools for implementing the results of analysis. A new subdiscipline, public- participation GIS, has grown up to study these issues, and to improve the use of GIS and spatial analysis in public decision making. Many of the arguments for the use of technology in support of decision making –for spatial decision support systems – centre on the benefits of these tools in settings that involve the potentially conflicting views of multiple stakeholders. Much has been written about spatial -analytic techniques that support multiple views, and address multiple criteria.

GIS may allow stakeholders to express their own views as sets of weights to be given to relevant factors. Saaty's Analytic Hierarchy Process is a widely used technique for eliciting such weights from stakeholders, and for deriving consensus weights and measures of agreement. Stakeholders benefit from the visualization capabilities of these systems, which allow them to see the effects of decisions in readily understood ways. They gain the impression that decisions are made scientifically, with abundant use of mathematics and computation, and are led to believe that these approaches represent a more objective, more desirable approach to debate and conflict resolution.

It is all too easy in such circumstances to see stakeholder satisfaction as the primary goal of the exercise. If stakeholders leave the room believing that a rigorous, scientific process has been conducted then everyone can feel that a useful exercise has come to an acceptable conclusion. None of these guarantees, however, that the results presented to the stakeholders are in fact based on good science. It is easy, with a little thought, to manipulate the outcomes of such processes to achieve hidden objectives. For example, when stakeholders are presented with five alternatives and asked to choose one, it is easy to see how the outcome might be manipulated by presenting a set that includes the desired outcome, plus four obviously unacceptable "red herrings". Experience suggests that stakeholders will find no difficulty in assigning relative measures of "importance" to factors, irrespective of whether the factors are or are not commensurate, and whether or not any definition of "importance" has been advanced and agreed.

Accuracy, Uncertainty and Cost

All measurements are subject to error, and science has developed sophisticated techniques for measuring instrument accuracy, and for determining how accuracy impacts the results of analysis. The basic principles of error analysis have been adapted to the specific needs of geographic data by Heuvelink and others, and statistical models have been developed for most of the standard geographic data types.

Uncertainty is often defined as the degree to which data leaves the user uncertain about the true nature of the real world. As such it presents a greater problem, because it derives not from errors in measurement, but from vagueness in definitions, lack of detail, and numerous other sources. When definitions are vague, there can be no objective definition of truth, but only the less satisfactory concept of consensus. A scientist steeped in traditional methodology would react by rejecting vague terms entirely, replacing them with terms that have rigorous definition, and are therefore capable of supporting replicability. Subjective terms such as "warm", "cold", "near", and "far" would be replaced by well-defined scales of temperature measurement and distance.

Nevertheless, GIS and to a lesser extent spatial analysis clearly exist at the interface between the rigorous, scientific world of well-defined terms and replicable experiments, and the vague, intuitive world of human discourse. Many users of GIS appear happy to work with vaguely defined classes of vegetation or land use, and there has been much interest in building user interfaces to GIS that come closer to emulating human ways of reasoning and discovering.

Naïve geography has been defined as a field that studies the simplifications humans often impose on the world around them, and writers have speculated about the potential for systems that also simplify –that "think more like humans do".

In the past decade or so there has been much interest in the application of fuzzy sets, rough sets, and related ideas in spatial analysis. There seems to be some degree of intuitive appeal in the idea of assigning degrees of membership to a class, even when the class is not itself well defined.

Methods have been devised for eliciting fuzzy membership values from professionals, from remotely sensed data, and from other sources, and for displaying these values in the form of maps. All of these methods stretch the norms of science, by arguing that it is possible to observe and measure useful properties despite a lack of agreement on the definitions of those properties. As such, they demand a re-examination of the basic tenets of scientific method.

Finally, spatial analysts find themselves today in a world overflowing with data. Satellite images, digital topographic maps, and a host of other sources provide an unprecedented opportunity for new and interesting research. Massive investments have been made over the past decade in data warehouses, spatial data centres, and geo-portals, with a view to facilitating the discovery and sharing of spatial data. Metadata standards have been devised that support search, by allowing researchers to hunt through catalogs looking for data that might meet their needs.

Yet almost certainly data discovered in this way will fail to meet the exact needs of the researcher. The data set will be too generalized, not sufficiently current, too inaccurate, or inadequate in another of a myriad of possible ways. In these circumstances it is inevitable that research objectives become modified to fit the properties of the available data, if the alternative is an exercise in field data collection that may be impossibly expensive. But the prevailing methodology of science says nothing about such compromises, maintaining instead that data must be exactly fit for purpose, and providing no basis on which users can find compromises between cost on the one hand, and accuracy or fitness for use on the other.

Modifiable Areal Unit Problem

The modifiable areal unit problem (MAUP) is a source of statistical bias that can radically affect the results of statistical hypothesis tests. It affects results when point-based measures of spatial phenomena (e.g. population density) are aggregated into districts. The resulting summary values (e.g. totals, rates, proportions) are influenced by the choice of district boundaries. For example, census data may be aggregated into census enumeration districts, or postcode areas, or police precincts, or *any* other spatial partition (thus, the "areal units" are "modifiable"). The issue was discovered in 1934 and later described in detail by Stan Openshaw, who lamented that "the areal units (zonal objects) used in many geographical studies are arbitrary, modifiable, and subject to the whims and fancies of whoever is doing, or did, the aggregating."

Statistical Bias

A statistic is biased if it is calculated in such a way that it is systematically different from the population parameter of interest. The following lists some types of biases, which can overlap. Selection bias involves individuals being more likely to be selected for study than others, biasing the sample. This can also be termed *Berksonian bias.*

Spectrum bias arises from evaluating diagnostic tests on biased patient samples, leading to an overestimate of the sensitivity and specificity of the test. The bias of an estimator is the difference between an estimator's expectations and the true value of the parameter being estimated. Omitted-variable bias is the bias that appears in estimates of parameters in a regression analysis when the assumed specification omits an independent variable that should be in the model.

In statistical hypothesis testing, a test is said to be unbiased when the probability of committing a type I error (i.e. false positive) is less than the significance level, and that of getting a true positive (rejecting the null hypothesis when the alternative hypothesis is true) is at least that of the significance level. Detection bias occurs when a phenomenon is more likely to be observed for a particular set of study subjects. For instance, the syndemic involving obesity and diabetes may mean doctors are more likely to look for diabetes in obese patients than in thinner patients, leading to an inflation in diabetes among obese patients because of skewed detection efforts.

Funding bias may lead to selection of outcomes, test samples, or test procedures that favour a study's financial sponsor. Reporting bias involves a skew in the availability of data, such that observations of a certain kind are more likely to be reported. Data-snooping bias comes from the misuse of data mining techniques. Analytical bias arise due to the way that the results are evaluated. Exclusion bias arise due to the systematic exclusion of certain individuals from the study.

Some Problems in Geospatial Information and GIS

The explosion of consumer demand for geospatial information and tools such as geographic information systems (GIS) to manipulate and graphically display such information has brought GIS into the daily lives of millions, whether they know it or not. Google Earth and handheld or dashboard navigation systems represent enormously popular examples of the wide variety of applications made possible through the availability of geospatial information.

The release of Google Earth in 2005 represented a paradigm shift in the way people understand geospatial information, according to some observers, because it offered multi-scale visualization of places and locations around the globe that was free and easy to use.

Historically, the federal government has been a primary provider of authoritative geospatial information, but some argue that consumer demand for spatial information has triggered a major shift toward local government and commercial providers.

The federal government has shifted, with some important exceptions, to consuming rather than providing geospatial information from a variety of sources. As a result, the federal government's role has shifted as well toward coordinating and managing geospatial data and facilitating partnerships among the producers and consumers of geospatial information in government, the private sector, and academia. The challenges to coordinating how geospatial data are acquired and used—collecting duplicative data sets, for example—at the local, state, and federal levels, in collaboration with the private sector, are long-standing and not yet resolved.

The explosion of geospatial data acquired at the local and state levels, for their own purposes and in conjunction with the private sector, underscores the long-recognised need for better coordination between the federal government and local and state authorities. At the same time, coordinating, managing, and facilitating the production and use of geospatial information from different sources, of different quality, and which was collected with specific objectives in mind has been a challenge.

GIS and Geospatial Data: A Primer

GIS is a computer system capable of capturing, storing, analyzing, and displaying geographically referenced information—information attached to a location, such as latitude and longitude, or street location.

Geographically referenced information is also known as geospatial information. Types of geospatial information include features like highway intersections, office buildings, rivers, the path of a tornado, the San Andreas Fault, or congressional district boundaries.

Information associated with a specific location is referred to in GIS parlance as an attribute, such as the population of a congressional district, or amount of movement per year along the San Andreas Fault.

The power of GIS is the ability to combine geospatial information in unique ways—by layers or themes—and extract something new. For instance, a GIS analysis might include the location of a highway intersection and the average number of vehicles that flow through the intersection throughout the day, and extract information useful for locating a business. GIS might include both the location of a river and the water depth along its course by season, and enable an analysis of the effects of development on runoff within the watershed. Overlaying the path of a severe thunderstorm with geospatial data on the types of structures encountered—homes, stores, schools, post offices—could inform an analysis of what types of building construction can survive high winds and hail.

Sources and Types of Geospatial Data

Geospatial data may be acquired by federal, state, tribal, county, and local governments, private companies, academic institutions, and nonprofit organisations. The collection and management of geospatial data are considered by many to be the costliest components of a GIS—some experts attribute close to 80% of GIS total costs to data acquisition.

It should be recognised that the amount of geospatial data is expanding rapidly, the methods for acquiring geospatial data are growing, and the ways geospatial data are being used is diversifying throughout local and state governments, as well as within the federal government. Geospatial data can be acquired using a variety of technologies. Land surveyors, census takers, aerial photographers, police, and even average citizens with a GPS-enabled cell phone can collect geospatial data using GPS or street addresses that can be entered into GIS.

The attributes of the collected data, such as land-use information, demographics, landscape features, or crime scene observations, can be entered manually or, in the case of a land survey map, digitized from a map format to a digital format by electronic scanning. Remote sensing data from satellites is acquired digitally and communicated to central facilities for processing and analysis in GIS. Digital satellite images, for example, can be analyzed in GIS to produce maps of land cover and land use.

When different types of geospatial data are combined in GIS (e.g., through combining satellite remote sensing land use information with aerial photograph data on housing development growth), the data must be transformed so they fit the same coordinates. GIS uses the processing power of a computer, together with geographic mapping techniques (cartography), to transform data from different sources onto one projection and one scale so that the data can be analyzed together.

Geospatial Data from Local, State, and Federal Governments and the Private Sector

Local and state governments provide geospatial data for use in GIS for a variety of public services such as land records, property taxation, local planning, subdivision control and zoning, and others.

Some observers note that local governments often contract with private sector companies to acquire more recent and higher-resolution data than what is available to the federal government.

Whether and how the most up-to-date and detailed geospatial information is made available to users other than the local government for whom the data were acquired are long-standing issues. For example, in the immediate aftermath of a natural disaster, it may be important for the federal government to acquire the most current and detailed geospatial information about the disaster area. In many instances, however, impediments to data sharing such as lack of interoperability between systems, restrictions on use, concerns about data security, and a lack of knowledge about what data exist and where the data can be found could hinder a timely and effective emergency response.

The federal government sometimes acquires or contracts to acquire geospatial data for federal needs, such as for updating floodplain maps from paper flood insurance rate maps to a digital format.

New techniques for collecting more data, such as Light Detection and Ranging (LIDAR), will help produce more accurate floodplain maps.

Example: Using LIDAR for Floodplain Mapping

There is no single nationwide elevation dataset of sufficient resolution and accuracy to make floodplain maps that meet FEMA requirements. A fundamental requirement for accurate flood maps is accurate elevation data, which are used to draw the boundaries for the 1-in-100 chance annual flood hazard (sometimes referred to as the 100-year flood). The USGS National Elevation

Dataset is a primary data source that FEMA uses to produce flood maps, but it has a level of uncertainty about 10 times larger than FEMA defines as acceptable for floodplain mapping.

The USGS National Elevation Dataset includes some high resolution, more accurate elevation data, but most of the USGS dataset is of poorer resolution. Alternate sources of more accurate elevation data exist, but are not available nationwide. One of these sources is provided using LIDAR, which can be used to collect high resolution elevation data Geospatial data are increasingly acquired and provided by the private sector, and many companies as well as professional organisations support and promote the role of private sector data providers.

GIS Layers or Themes

The attributes of different types of geospatial data—such as land ownership, roads and bridges, buildings, lakes and rivers, counties, or congressional districts—can each constitute a layer or theme in GIS. GIS has the ability to link and integrate information from several different data layers or themes over the same geographic coordinates, which is very difficult to do with any other means. For example, GIS could combine a major road from one data layer as the boundary dividing land zoned for commercial development with the location of wetlands from another data layer. Precipitation data, from a third data layer, could be combined with a fourth data layer that shows streams and rivers. GIS could then be used to calculate where and how much runoff might flow from the commercial development into the wetlands. Thus the power of GIS analysis can be used to create a new way to interpret information that would otherwise be very difficult to visualize and analyze.

Current Options and Software Assessment

Perhaps the first question asked by anyone when discovering GIS is *what are the current options available ?*. This question is often asked as directly as *what is the best GIS ?*. Quite simply, there is no best GIS. A wide variety of GIS software offerings exist in the commercial market place. Commercial surveys often are a good starting point in the assessment of GIS software. The number of GIS software offerings is approximately 10 if one eliminates the following:

➡	the university based research software, which tends to lack full integration and usually has narrow channels of functionality;
➡	the CAD vendors, who like to use GIS jargon but often cannot provide full featured functionality; and
➡	the consulting firms, that will provide or customize selected modules for a GIS but lack a complete product.

One of the problems in evaluating the functionality of GIS software is the bias one gets from using one system or another. Comparing similar functions between systems is often confusing. Like any software, ultimately some do particular tasks better than others, and also some lack functionality compared to others.

Due mostly to this diverse range of different architectures and the complex nature of spatial analysis no standard evaluation technique or method has been established to date. Any GIS should be evaluated strictly in terms of the potential user's needs and requirements in consideration of their work procedures, production requirements, and organisational context ! The experienced GIS consultant can play a large and valuable role in the assessment process.

A current accepted approach to selecting the appropriate GIS involves establishing a benchmark utilizing real data that best represents the normal workflow and processes employed in your organisation. The identification of potential needs and requirements is essential in developing a proper benchmark with which to evaluate GIS software packages. A formalized user need analysis is absolutely critical to the successful implementation of GIS technology.

Development of the benchmark should include a consideration of other roles within your organisation that may require integration with the GIS technology. A logical and systematic approach as such is consistent with existing information systems (IS) planning methodologies and will ultimately provide a mechanism for a successful evaluation process.

Implementation Issues

The mere presence of an implementation plan does not guarantee success. Most organisations do not have sufficient staff to cope with the commitment and extra work required when introducing a GIS to existing operations. GIS implementation must also consider all technology transfer processes.

Common Pitfalls

Several pitfalls exist that most often contribute to the failure of a GIS implementation strategy. These are identified below:

➧	Failure to identify and involve all users Users in an operational GIS environment consist of operations, management, and policy levels of the organization. All three levels should be considered when identifying the needs of your users.
➧	Failure to match GIS capability and needs. A wide spectrum of GIS hardware and software choices currently exist. The buyer is presented with a significant challenge making the right choice. Remember, the right choice will be the GIS that provides the needed performance no more, no less for the minimum investment. The success of a GIS implementation is particularly sensitive to the right hardware and software choices !
➧	Failure to identify total costs. The GIS acquisition cost is relatively easy to identify. However, it will represent a very small fraction of the total cost of implementing a GIS. Ongoing costs are substantial and include hardware and software maintenance, staffing, system administration, initial data loading, data updating, custom programming, and consulting fees.
➧	Failure to conduct a pilot study The GIS implementation plan concerns itself with the many technical and administrative issues and their related cost impacts. Three of the most crucial issues, are database design, data loading and maintenance, and day-to-day operations. The pilot study will allow you to gather detailed observations, provided it is properly designed, to allow you to effectively estimate the operational requirements.
➧	Giving the GIS implementation responsibility to the EDP Department. Because of the distinct differences of the GIS from conventional EDP systems, the GIS implementation team is best staffed by non-data processing types. The specialized skills of the 'GIS analyst' are required at this stage. Reliance on conventional EDP personnel who lack these skills will ensure failure.
➧	Failure to consider technology transfer. Training and support for on-going learning, for in-house staff as well as new personnel, is essential for a successful implementation. Staff at the three levels should be educated with respect to the role of the GIS in the organization. Education and knowledge of the GIS can only be obtained through on-going learning exercises. Nothing can replace the investment of hands on time with a GIS !

The Learning Curve

Contrary to information provided by commercial vendors of GIS software, there is a substantial learning curve associated with GIS. It is normally not a technology that one becomes proficient in overnight. It requires an understanding of geographical relationships accompanied by committed hands-on time to fully apply the technology in a responsible and cost effective manner. Proficiency and productivity are only obtained through applied hands on with the system ! GIS is an applied science. Nothing can replace the investment of *hands-on* with GIS. The following figure presents the typical learning curve for GIS installations.

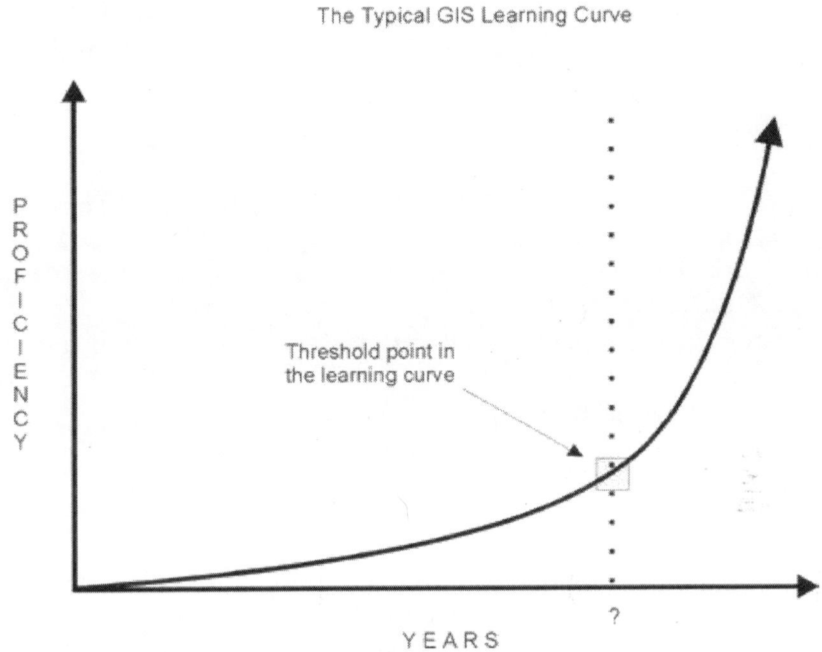

Figure: *The learning curve is dependent on a variety of factors including:*

➡	the amount of time spent by the individual with hands-on access;
➡	the skills, aptitude and motivation of the individual;
➡	the commitment and priority attached to GIS technology dictated by the organization and management;
➡	the availability of data; and
➡	the choice of software and hardware platforms.

A critical requirement for all GIS implementations is that adequate education and training is provided for operational staff, as well as realistic priorities are defined with which to learn and apply the technology. This is where a formal training curriculum is required to ensure that time is dedicated to learning the technology properly. Adding GIS activities to a staff member's responsibilities without establishing well defined milestones and providing adequate time and training mechanisms is prone to failure. A focused and properly trained operations staff that has consistent training will result in greatly reduced turnaround times for operations, and ensure consistency in quality of product. The threshold point of the learning curve is typically around the two year time frame. However, this is dependent on the ability of the organisation to establish a well defined and structured implementation plan that affords appropriate training and resources for technical staff. The flat part of the learning curve can be shortened if proper training is provided, data is available for use, the *right* software and hardware is acquired.

The typical learning curve reflects a long initial period for understanding spatial data compilation requirements and database architecture. However, after data models are well understood and sufficient data compilation has been completed the learning curve accelerates. Once a formal application development environment is established and user needs are well defined an infrastructure exists for effective application of the technology. Building operational applications based on formal functional specifications will result in continued accelerated learning. The *data hurdle* is often a stumbling block for many GIS users.

The Productivity Curve

GIS is a long term investment that matures over time. The turnaround for results may be longer than initially expected. The establishment of a formal implementation strategy will help to ensure that realistic expectations are met. Data is the framework for successful application of GIS technology. In this respect, the investment in establishing a solid data platform will reap rewards in a short term timeframe for establishing a cost-effective and productive GIS operation. The availability of quality data supplemented by a planned implementation strategy are the cornerstones of achieving a productive and successful GIS operation. A robust database should be considered an asset !

However, even with a well defined and systematic implementation strategy GIS technology will not provide immediate benefits. Benefits and increased productivity are not achieved overnight. GIS technology is complex in nature, has a generally steep learning curve, and requires a *complement of skills* for it to be applied successfully. In fact, most organisations realise a loss in overall operational productivity over the short term while the GIS platforms are being installed, staff is trained, the learning curve is initiated, and data is being captured. This is common of all office automation activities. The following figure presents the typical situation that occurs with respect to comparing long term productivity with, and without, GIS technology.

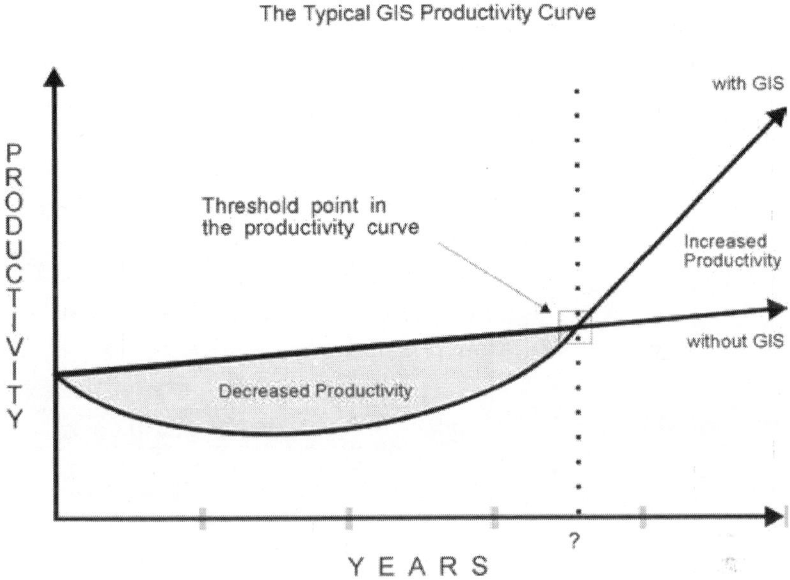

Figure: *This graph represents the typical situation that occurs with respect to comparing long term productivity with, and without, GIS technology.*

Depending on the unique circumstances of the implementation process, the status of data compilation, and the organisational climate, increased productivity is normally reached between the second and fifth year of implementation. This is identified by the threshold point. Again, this is dependent on a variety of factors including:

➡	the skills and experience of the staff involved;
➡	the priority and commitment by the organization;
➡	the implementation strategy; and
➡	the status of data compilation.

The primary issue with implementing GIS is to achieve the threshold point of increased productivity in the shortest possible time frame. In other words, minimize the time in which a decrease in productivity occurs. Of course, the issue of productivity is typically of greaterst concern with private industry, e.g. forestry companies. Nonetheless, the significant investment in hardware/software, data, and training necessitates that a structured approach be utilized to achieve the threshold point in the shortest possible time frame.

A GIS acquisition based on well defined user needs and priorities is more likely to succeed than without. A major pitfall of most installations with GIS technology, e.g. particularly forestry companies and government agencies, is the lack of well defined user needs on which to base the GIS acquisition and implementation.

The Implementation Plan

Implementation can be seen as a six phase process. They are: Creating an awareness GIS needs to be *sold* within an organisation. The education of staff is very important. Depending on the way in which GIS technology is being introduced to the organisation the process for creating an awareness may differ. Technical workshops are often appropriate when a *top-down* approach exists, while management workshops are often more relevant when a *bottoms-up* approach exists. Education of the new technology should focus on identifying existing problems within an organisation. These often help justify a GIS acquisition. They include :

➡	spatial information is poorly maintained or out of date;
➡	spatial data is not recorded or stored in a standard way;
➡	spatial data may not be defined in a consistent manner, e.g. different classifications for timber information;
➡	data is not shared between departments within an organization;
➡	data retrieval and manipulation capabilities are inadequate to meet existing needs; and
➡	new demands are made on the organization that cannot be met with existing information systems.
➡	Identifying System Requirements

The definition of system requirements is usually done in a *user needs analysis*. A user needs analysis identifies users of a system and all information products required by those users. Often a prioritization of the information products and the data requirements of those products is also undertaken. A proper user needs analysis is crucial to the successful evaluation of GIS software alternatives.

After user needs have been identified and prioritized they must be translated into *functional requirements*. Ideally, the functional requirements definition will result in a set of processing functions, system capabilities, and hardware requirements, e.g. data storage, performance. Experienced GIS consultants often play a major role in this phase.

System Evaluations

Evaluating alternative hardware and software solutions is normally conducted in several stages. Initially a number of candidate systems are identified. Information to support this process is acquired through demonstrations, vendor literature, etc. A short listing of candidates normally occurs based on a low level assessment. This followed by a high level assessment based on the functional requirements identified in the previous phase. This often results in a rating matrix or template. The assessment should take into account production priorities and their appropriate functional translation. After systems have been evaluated based on functional requirements a short list is prepared for those vendors deemed suitable. A standard benchmark, as discussed earlier, is then used to determine the system of choice.

Justifying the System Acquisition

The proper justification of the chosen system requires consideration of several factors. Typically a *cost-benefit analysis* is undertaken to analyze the expected costs and benefits of acquiring a system. To proceed further with acquisition the GIS should provide considerable benefits over expected costs. It is important that the identification of intangible benefits also be considered. The justification process should also include an evaluation of other requirements. These include data base development requirements, e.g. existing data versus new data needs and associated costs; technological needs, e.g. maintenance, training, and organisational requirements, e.g. new staff, reclassification of existing job descriptions for those staff who will use the GIS.

System Acquisition and Start Up

After the system, e.g. hardware, software, and data, is acquired the start up phase begins. This phase should include *pilot projects.* Pilot projects are a valuable means of assessing progress and identifying problems early, before significant resources have been wasted. Also, because of the costs associated with implementing a GIS it is often appropriate to generate some results quickly to appease management. First impressions are often long remembered.

Operational Phase

The operational phase of a GIS implementation involves the on-going maintenance, application, and development of the GIS. The issue of responsibility for the system and liability is critical. It is important that appropriate security and transaction control mechanisms be in place to support the system. A systematic approach to system management, e.g. hardware, software, and data, is essential.

Global Positioning System and Geography

The Global Positioning System (GPS) is a space-based satellite navigation system that provides location and time information in all weather conditions, anywhere on or near the Earth where there is an unobstructed line of sight to four or more GPS satellites. The system provides critical capabilities to military, civil and commercial users around the world. It is maintained by the United States government and is freely accessible to anyone with a GPS receiver.

The GPS project was developed in 1973 to overcome the limitations of previous navigation systems, integrating ideas from several predecessors, including a number of classified engineering design studies from the 1960s. GPS was created and realised by the U.S. Department of Defence (DoD) and was originally run with 24 satellites. It became fully operational in 1995. Bradford Parkinson, Roger L. Easton, and Ivan A. Getting are credited with inventing it. Advances in technology and new demands on the existing system have now led to efforts to modernise the GPS system and implement the next generation of GPS III satellites and Next Generation Operational Control System (OCX). Announcements from Vice President Al Gore and the White House in 1998 initiated these changes. In 2000, the U.S. Congress authorized the modernisation effort, GPS III.

In addition to GPS, other systems are in use or under development. The Russian Global Navigation Satellite System (GLONASS) was developed contemporaneously with GPS, but suffered from incomplete coverage of the globe until the mid-2000s. There are also the planned European Union Galileo positioning system, India's Indian Regional Navigation Satellite System, and the Chinese Beidou Navigation Satellite System.

History

The design of GPS is based partly on similar ground-based radio-navigation systems, such as LORAN and the Decca Navigator, developed in the early 1940s and used by the British Royal Navy during World War II.

Predecessors

In 1956, the German-American physicist Friedwardt Winterberg proposed a test of general relativity - detecting time slowing in a strong gravitational field using accurate atomic clocks placed in orbit inside artificial satellites. Calculations using general relativity determined that the clocks on the GPS satellites would be seen by Earth's observers to run 38 microseconds faster per day (than those on Earth), and this was corrected for in the design of GPS.

The Soviet Union launched the first man-made satellite, Sputnik, in 1957. Two American physicists, William Guier and George

Weiffenbach, at Johns Hopkins's Applied Physics Laboratory (APL), decided to monitor Sputnik's radio transmissions. Within hours they realised that, because of the Doppler effect, they could pinpoint where the satellite was along its orbit. The Director of the APL gave them access to their UNIVAC to do the heavy calculations required. The next spring, Frank McClure, the deputy director of the APL, asked Guier and Weiffenbach to investigate the inverse problem— pinpointing the user's location given that of the satellite. (At the time, the Navy was developing the submarine-launched Polaris missile, which required them to know the submarine's location.) This led them and APL to develop the Transit system. In 1959, ARPA (renamed DARPA in 1972) also played a role in Transit.

The first satellite navigation system, Transit, used by the United States Navy, was first successfully tested in 1960. It used a constellation of five satellites and could provide a navigational fix approximately once per hour. In 1967, the U.S. Navy developed the Timation satellite that proved the ability to place accurate clocks in space, a technology required by GPS. In the 1970s, the ground-based Omega Navigation System, based on phase comparison of signal transmission from pairs of stations, became the first worldwide radio navigation system. Limitations of these systems drove the need for a more universal navigation solution with greater accuracy.

While there were wide needs for accurate navigation in military and civilian sectors, almost none of those was seen as justification for the billions of dollars it would cost in research, development, deployment, and operation for a constellation of navigation satellites. During the Cold War arms race, the nuclear threat to the existence of the United States was the one need that did justify this cost in the view of the United States Congress. This deterrent effect is why GPS was funded. It is also the reason for the ultra secrecy at that time. The nuclear triad consisted of the United States Navy's submarine-launched ballistic missiles (SLBMs) along with United States Air Force (USAF) strategic bombers and intercontinental ballistic missiles (ICBMs). Considered vital to the nuclear-deterrence posture, accurate determination of the SLBM launch position was a force multiplier.

Precise navigation would enable United States submarines to get an accurate fix of their positions before they launched their SLBMs. The USAF, with two thirds of the nuclear triad, also had requirements for a more accurate and reliable navigation system. The Navy and Air Force were developing their own technologies in parallel to solve what was essentially the same problem. To increase the survivability of ICBMs, there was a proposal to use mobile launch platforms (such as

Russian SS-24 and SS-25) and so the need to fix the launch position had similarity to the SLBM situation.

In 1960, the Air Force proposed a radio-navigation system called MOSAIC (MObile System for Accurate ICBM Control) that was essentially a 3-D LORAN. A follow-on study, Project 57, was worked in 1963 and it was "in this study that the GPS concept was born". That same year, the concept was pursued as Project 621B, which had "many of the attributes that you now see in GPS" and promised increased accuracy for Air Force bombers as well as ICBMs. Updates from the Navy Transit system were too slow for the high speeds of Air Force operation. The Naval Research Laboratory continued advancements with their Timation (Time Navigation) satellites, first launched in 1967, and with the third one in 1974 carrying the first atomic clock into orbit.

Another important predecessor to GPS came from a different branch of the United States military. In 1964, the United States Army orbited its first Sequential Collation of Range (SECOR) satellite used for geodetic surveying. The SECOR system included three ground-based transmitters from known locations that would send signals to the satellite transponder in orbit. A fourth ground-based station, at an undetermined position, could then use those signals to fix its location precisely. The last SECOR satellite was launched in 1969. Decades later, during the early years of GPS, civilian surveying became one of the first fields to make use of the new technology, because surveyors could reap benefits of signals from the less-than-complete GPS constellation years before it was declared operational. GPS can be thought of as an evolution of the SECOR system where the ground-based transmitters have been migrated into orbit.

Development

With these parallel developments in the 1960s, it was realised that a superior system could be developed by synthesizing the best technologies from 621B, Transit, Timation, and SECOR in a multi-service program.

During Labour Day weekend in 1973, a meeting of about twelve military officers at the Pentagon discussed the creation of a *Defence Navigation Satellite System (DNSS)*. It was at this meeting that "the real synthesis that became GPS was created." Later that year, the DNSS program was named *Navstar*, or Navigation System Using Timing and Ranging. With the individual satellites being associated with the name Navstar (as with the predecessors Transit and Timation), a more fully encompassing name was used to identify the constellation of Navstar satellites, *Navstar-GPS*, which was later shortened simply to GPS. Ten "Block I" prototype satellites were launched between 1978 and 1985 (with one prototype being destroyed in a launch failure).

After Korean Air Lines Flight 007, a Boeing 747 carrying 269 people, was shot down in 1983 after straying into the USSR's prohibited airspace, in the vicinity of Sakhalin and Moneron Islands, President Ronald Reagan issued a directive making GPS freely available for civilian use, once it was sufficiently developed, as a common good. The first satellite was launched in 1989, and the 24th satellite was launched in 1994. The GPS program cost at this point, not including the cost of the user equipment, but including the costs of the satellite launches, has been estimated to be about USD$5 billion (then-year dollars). Roger L. Easton is widely credited as the primary inventor of GPS.

Initially, the highest quality signal was reserved for military use, and the signal available for civilian use was intentionally degraded (Selective Availability). This changed with President Bill Clinton ordering Selective Availability to be turned off at midnight May 1, 2000, improving the precision of civilian GPS from 100 to 20 metres (328 to 66 ft). The executive order signed in 1996 to turn off Selective Availability in 2000 was proposed by the U.S. Secretary of Defence, William Perry, because of the widespread growth of differential GPS services to improve civilian accuracy and eliminate the U.S. military advantage. Moreover, the U.S. military was actively developing technologies to deny GPS service to potential adversaries on a regional basis.

Over the last decade, the U.S. has implemented several improvements to the GPS service, including new signals for civil use and increased accuracy and integrity for all users, all while maintaining compatibility with existing GPS equipment.

GPS modernisation has now become an ongoing initiative to upgrade the Global Positioning System with new capabilities to meet growing military, civil, and commercial needs. The program is being implemented through a series of satellite acquisitions, including GPS Block III and the Next Generation Operational Control System (OCX). The U.S. Government continues to improve the GPS space and ground segments to increase performance and accuracy.

GPS is owned and operated by the United States Government as a national resource. Department of Defence (DoD) is the steward of GPS. *Interagency GPS Executive Board (IGEB)* oversaw GPS policy matters from 1996 to 2004. After that the National Space-Based Positioning, Navigation and Timing Executive Committee was established by presidential directive in 2004 to advise and coordinate federal departments and agencies on matters concerning the GPS and related systems. The executive committee is chaired jointly by the deputy secretaries of defence and transportation. Its membership includes equivalent-level officials from the departments of state, commerce, and homeland security, the joint chiefs of staff, and NASA. Components of the executive office of the president participate as observers to the executive committee, and the FCC chairman participates as a liaison.

The DoD is required by law to "maintain a Standard Positioning Service (as defined in the federal radio navigation plan and the standard positioning service signal specification) that will be available on a continuous, worldwide basis," and "develop measures to prevent hostile use of GPS and its augmentations without unduly disrupting or degrading civilian uses."

Basic Concept of GPS

Fundamentals

The GPS system concept is based on *time.* The satellites carry atomic clocks which are synchronized and very stable; any drift from true time maintained on the ground is corrected daily. Likewise, the satellite locations are monitored precisely. User receivers have clocks as well. However, they are not synchronized with true time, and are less stable. GPS satellites transmit data continuously which contains their current time and position. A GPS receiver listens to multiple satellites and solves equations to determine the exact position of the receiver and its deviation from true time. At a minimum, four satellites must be in view of the receiver in order to compute four unknown quantities (three position coordinates and clock deviation from satellite time).

More Detailed Description

Each GPS satellite continually broadcasts a signal (carrier frequency with modulation) that includes: a pseudorandom code (sequence of ones and zeros) that is known to the receiver. By time-aligning a receiver-generated version and the receiver-measured version of the code, the time of arrival (TOA) of a defined point in the code sequence, called an epoch, can be found in the receiver clock time scale a message that includes the time of transmission (TOT) of the code epoch (in GPS system time scale) and the satellite position at that time.

Conceptually, the receiver measures the TOAs (according to its own clock) of four satellite signals. From the TOAs and the TOTs, the receiver forms four time of flight (TOF) values, which are (given the speed of light) approximately equivalent to receiver-satellite range differences. The receiver then computes its three-dimensional position and clock deviation from the four TOFs.

In practice the receiver position (in three dimensional Cartesian coordinates with origin at the earth's centre) and the offset of the receiver clock relative to GPS system time are computed simultaneously, using the navigation equations to process the TOFs.

The receiver's earth-centred solution location is usually converted to latitude, longitude and height relative to an ellipsoidal earth model. The height may then be further converted to height relative the geoid (e.g., EGM96) (essentially, mean sea level). These coordinates may be displayed, perhaps on a moving map display and/or recorded and/or used by other system (e.g., vehicle guidance).

User-Satellite Geometry

Although usually not formed explicitly in the receiver processing, the conceptual TDOAs define the measurement geometry. Each TDOA corresponds to a hyperboloid of revolution. The line connecting the two satellites involved (and its extensions) forms the axis of the hyperboloid. The receiver is located at the point where three hyperboloids intersect. It is sometimes incorrectly said that the user location is at the intersection of three spheres. While simpler to visualize, this is only the case if the receiver has a clock synchronized with the satellite clocks (i.e., the receiver measures true ranges to the satellites rather than range differences). There are significant performance benefits to the user carrying a clock synchronized with the satellites. Foremost is that only three satellites are needed to compute a position solution. If this were part of the GPS system concept so that all users needed to carry a synchronized clock, then a smaller number of satellites could be deployed. However, the cost and complexity of the user equipment would increase significantly.

Receiver in Continuous Operation

The description above is representative of a receiver start-up situation. Most receivers have a track algorithm, sometimes called a *tracker*, that combines sets of satellite measurements collected at different times — in effect, taking advantage of the fact that successive receiver positions are usually close to each other. After a set of measurements are processed, the tracker predicts the receiver location corresponding to the next set of satellite measurements. When the new measurements are collected, the receiver uses a weighting scheme to combine the new measurements with the tracker prediction. In general, a tracker can (a) improve receiver position and time accuracy,

(b) reject bad measurements, and (c) estimate receiver speed and direction. The disadvantage of a tracker is that changes in speed or direction can only be computed with a delay, and that derived direction becomes inaccurate when the distance travelled between two position measurements drops below or near the random error of position measurement. GPS units can use measurements of the doppler shift of the signals received to compute velocity accurately. More advanced navigation systems use additional sensors like a compass or an inertial navigation system to complement GPS.

Non-navigation Applications

In typical GPS operation as a navigator, four or more satellites must be visible to obtain an accurate result. The solution of the navigation equations gives the position of the receiver along with the difference between the time kept by the receiver's on-board clock and the true time-of-day, thereby eliminating the need for a more precise and possibly impractical receiver based clock. Applications for GPS such as time transfer, traffic signal timing, and synchronization of cell phone base stations, make use of this cheap and highly accurate timing. Some GPS applications use this time for display, or, other than for the basic position calculations, do not use it at all.

Although four satellites are required for normal operation, fewer apply in special cases. If one variable is already known, a receiver can determine its position using only three satellites. For example, a ship or aircraft may have known elevation. Some GPS receivers may use additional clues or assumptions such as reusing the last known altitude, dead reckoning, inertial navigation, or including information from the vehicle computer, to give a (possibly degraded) position when fewer than four satellites are visible.

Structure

The current GPS consists of three major segments. These are the space segment (SS), a control segment (CS), and a user segment (US). The U.S. Air Force develops, maintains, and operates the space and control segments. GPS satellites broadcast signals from space, and each GPS receiver uses these signals to calculate its three-dimensional location (latitude, longitude, and altitude) and the current time.

The space segment is composed of 24 to 32 satellites in medium Earth orbit and also includes the payload adapters to the boosters required to launch them into orbit. The control segment is composed of a master control station (MCS), an alternate master control station, and a host of dedicated and shared ground antennas and monitor stations. The user segment is composed of hundreds of thousands of U.S. and allied military users of the secure GPS Precise Positioning Service, and tens of millions of civil, commercial, and scientific users of the Standard Positioning Service.

Space Segment

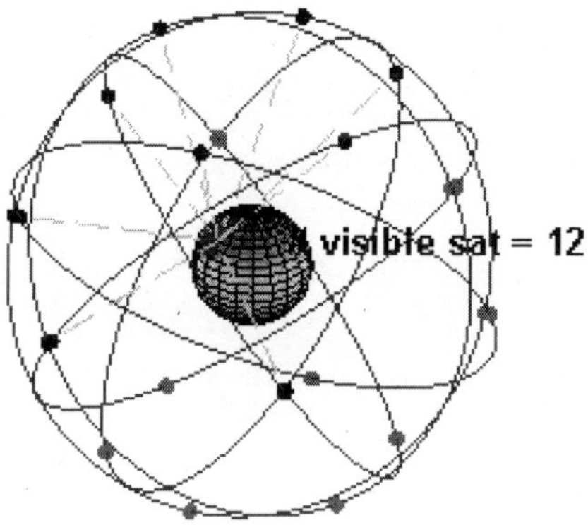

Figure: *A visual example of a 24 satellite GPS constellation in motion with the Earth rotating. Notice how the number of* satellites in view *from a given point on the Earth's surface, in this example at 45°N, changes with time.*

The space segment (SS) is composed of the orbiting GPS satellites, or Space Vehicles (SV) in GPS parlance. The GPS design originally called for 24 SVs, eight each in three approximately circular orbits, but this was modified to six orbital planes with four satellites each. The six orbit planes have approximately 55° inclination (tilt relative to Earth's equator) and are separated by 60° right ascension of the ascending node (angle along the equator from a reference point to the orbit's intersection). The orbital period is one-half a sidereal day, i.e., 11 hours and 58 minutes so that the satellites pass over the same locations or almost the same locations every day. The orbits are arranged so that at least six satellites are always within line of sight from almost everywhere on Earth's surface. The result of this objective is that the four satellites are not evenly spaced (90 degrees) apart within each orbit. In general terms, the angular difference between satellites in each orbit is 30, 105, 120, and 105 degrees apart which sum to 360 degrees.

Orbiting at an altitude of approximately 20,200 km (12,600 mi); orbital radius of approximately 26,600 km (16,500 mi), each SV makes two complete orbits each sidereal day, repeating the same ground track each day. This was very helpful during development because even with only four satellites, correct alignment means all four are visible from one spot for a few hours each day. For military operations, the ground track repeat can be used to ensure good coverage in combat zones.

As of December 2012, there are 32 satellites in the GPS constellation. The additional satellites improve the precision of GPS receiver calculations by providing redundant measurements. With the increased number of satellites, the constellation was changed to a nonuniform arrangement. Such an arrangement was shown to improve reliability and availability of the system, relative to a uniform system, when multiple satellites fail. About nine satellites are visible from any point on the ground at any one time, ensuring considerable redundancy over the minimum four satellites needed for a position.

Control Segment

The control segment is composed of:

a master control station (MCS),

an alternate master control station,

four dedicated ground antennas, and

six dedicated monitor stations.

The MCS can also access U.S. Air Force Satellite Control Network (AFSCN) ground antennas (for additional command and control capability) and NGA (National Geospatial-Intelligence Agency) monitor stations. The flight paths of the satellites are tracked by dedicated U.S. Air Force monitoring stations in Hawaii, Kwajalein Atoll, Ascension Island, Diego Garcia, Colorado Springs, Colorado and Cape Canaveral, along with shared NGA monitor stations operated in England, Argentina, Ecuador, Bahrain, Australia and Washington DC. The tracking information is sent to the Air Force Space Command MCS at Schriever Air Force Base 25 km (16 mi) ESE of Colorado Springs, which is operated by the 2nd Space Operations Squadron (2 SOPS) of the U.S. Air Force. Then 2 SOPS contacts each GPS satellite regularly with a navigational update using dedicated or shared (AFSCN) ground antennas (GPS dedicated ground antennas are located at Kwajalein, Ascension Island, Diego Garcia, and Cape Canaveral). These updates synchronize the atomic clocks on board the satellites to within a few nanoseconds of each other, and adjust the ephemeris of each satellite's internal orbital model. The updates are created by a Kalman filter that uses inputs from the ground monitoring stations, space weather information, and various other inputs.

Satellite maneuvers are not precise by GPS standards. So to change the orbit of a satellite, the satellite must be marked *unhealthy*, so receivers will not use it in their calculation. Then the maneuver can be carried out, and the resulting orbit tracked from the ground. Then the new ephemeris is uploaded and the satellite marked healthy again.

The Operation Control Segment (OCS) currently serves as the control segment of record. It provides the operational capability that supports global GPS users and keeps the GPS system operational and performing within specification.

OCS successfully replaced the legacy 1970s-era mainframe computer at Schriever Air Force Base in September 2007. After installation, the system helped enable upgrades and provide a foundation for a new security architecture that supported the U.S. armed forces. OCS will continue to be the ground control system of record until the new segment, Next Generation GPS Operation Control System (OCX), is fully developed and functional.

The new capabilities provided by OCX will be the cornerstone for revolutionizing GPS's mission capabilities, and enabling Air Force Space Command to greatly enhance GPS operational services to U.S. combat forces, civil partners and myriad domestic and international users.

The GPS OCX program also will reduce cost, schedule and technical risk. It is designed to provide 50% sustainment cost savings through efficient software architecture and Performance-Based Logistics. In addition, GPS OCX expected to cost millions less than the cost to upgrade OCS while providing four times the capability.

The GPS OCX program represents a critical part of GPS modernisation and provides significant information assurance improvements over the current GPS OCS program.

- OCX will have the ability to control and manage GPS legacy satellites as well as the next generation of GPS III satellites, while enabling the full array of military signals.
- Built on a flexible architecture that can rapidly adapt to the changing needs of today's and future GPS users allowing immediate access to GPS data and constellations status through secure, accurate and reliable information.
- Empowers the warfighter with more secure, actionable and predictive information to enhance situational awareness.
- Enables new modernised signals (L1C, L2C, and L5) and has M-code capability, which the legacy system is unable to do.
- Provides significant information assurance improvements over the current program including detecting and preventing cyber attacks, while isolating, containing and operating during such attacks.
- Supports higher volume near real-time command and control capabilities and abilities.

On September 14, 2011, the U.S. Air Force announced the completion of GPS OCX Preliminary Design Review and confirmed that the OCX program is ready for the next phase of development.

The GPS OCX program has achieved major milestones and is on track to support the GPS IIIA launch in May 2014.

User Segment

The user segment is composed of hundreds of thousands of U.S. and allied military users of the secure GPS Precise Positioning Service, and tens of millions of civil, commercial and scientific users of the Standard Positioning Service. In general, GPS receivers are composed of an antenna, tuned to the frequencies transmitted by the satellites, receiver-processors, and a highly stable clock (often a crystal oscillator). They may also include a display for providing location and speed information to the user. A receiver is often described by its number of channels: this signifies how many satellites it can monitor simultaneously. Originally limited to four or five, this has progressively increased over the years so that, as of 2007, receivers typically have between 12 and 20 channels.

GPS receivers may include an input for differential corrections, using the RTCM SC-104 format. This is typically in the form of an RS-232 port at 4,800 bit/s speed. Data is actually sent at a much lower rate, which limits the accuracy of the signal sent using RTCM. Receivers with internal DGPS receivers can outperform those using external RTCM data. As of 2006, even low-cost units commonly include Wide Area Augmentation System (WAAS) receivers.

A typical GPS receiver with integrated antenna.

Many GPS receivers can relay position data to a PC or other device using the NMEA 0183 protocol. Although this protocol is officially defined by the National Marine Electronics Association (NMEA), references to this protocol have been compiled from public records, allowing open source tools like gpsd to read the protocol without violating intellectual property laws. Other proprietary protocols exist as well, such as the SiRF and MTK protocols. Receivers can interface with other devices using methods including a serial connection, USB, or Bluetooth.

Applications

While originally a military project, GPS is considered a *dual-use* technology, meaning it has significant military and civilian applications.

GPS has become a widely deployed and useful tool for commerce, scientific uses, tracking, and surveillance. GPS's accurate time facilitates everyday activities such as banking, mobile phone operations, and even the control of power grids by allowing well synchronized hand-off switching.

Civilian

Many civilian applications use one or more of GPS's three basic components: absolute location, relative movement, and time transfer.

- Astronomy: both positional and clock synchronization data is used in Astrometry and Celestial mechanics calculations. It is also used in amateur astronomy using small telescopes to professionals observatories, for example, while finding extrasolar planets.
- Automated vehicle: applying location and routes for cars and trucks to function without a human driver.
- Cartography: both civilian and military cartographers use GPS extensively.
- Cellular telephony: clock synchronization enables time transfer, which is critical for synchronizing its spreading codes with other base stations to facilitate inter-cell handoff and support hybrid GPS/cellular position detection for mobile emergency calls and other applications. The first handsets with integrated GPS launched in the late 1990s. The U.S. Federal Communications Commission (FCC) mandated the feature in either the handset or in the towers (for use in triangulation) in 2002 so emergency services could locate 911 callers. Third-party software developers later gained access to GPS APIs from Nextel upon launch, followed by Sprint in 2006, and Verizon soon thereafter.
- Clock synchronization: the accuracy of GPS time signals (±10 ns) is second only to the atomic clocks upon which they are based.
- Disaster relief/emergency services: depend upon GPS for location and timing capabilities.
- Meteorology-Upper Airs: measure and calculate the atmospheric pressure, wind speed and direction up to 27 km from the earth's surface
- Fleet Tracking: the use of GPS technology to identify, locate and maintain contact reports with one or more fleet vehicles in real-time.
- Geofencing: vehicle tracking systems, person tracking systems, and pet tracking systems use GPS to locate a vehicle, person, or pet. These devices are attached to the vehicle, person, or the pet collar. The application provides continuous tracking and

mobile or Internet updates should the target leave a designated area.

- Geotagging: applying location coordinates to digital objects such as photographs (in Exif data) and other documents for purposes such as creating map overlays with devices like Nikon GP-1
- GPS Aircraft Tracking
- GPS for Mining: the use of RTK GPS has significantly improved several mining operations such as drilling, shoveling, vehicle tracking, and surveying. RTK GPS provides centimetre-level positioning accuracy.
- GPS tours: location determines what content to display; for instance, information about an approaching point of interest.
- Navigation: navigators value digitally precise velocity and orientation measurements.
- Phasor measurements: GPS enables highly accurate timestamping of power system measurements, making it possible to compute phasors.
- Recreation: for example, geocaching, geodashing, GPS drawing and waymarking.
- Robotics: self-navigating, autonomous robots using a GPS sensors, which calculate latitude, longitude, time, speed, and heading.
- Sport: used in football and rugby for the control and analysis of the training load.
- Surveying: surveyors use absolute locations to make maps and determine property boundaries.
- Tectonics: GPS enables direct fault motion measurement of earthquakes. Between earthquakes GPS can be used to measure crustal motion and deformation to estimate seismic strain buildup for creating seismic hazard maps.
- Telematics: GPS technology integrated with computers and mobile communications technology in automotive navigation systems

Restrictions on Civilian Use

The U.S. government controls the export of some civilian receivers. All GPS receivers capable of functioning above 18 kilometres (11 mi) altitude and 515 metres per second (1,001 kn) or designed, modified for use with unmanned air vehicles like e.g. ballistic or cruise missile systems are classified as munitions (weapons) for which State

Department export licenses are required. This rule applies even to otherwise purely civilian units that only receive the L1 frequency and the C/A (Coarse/Acquisition) code.

Disabling operation above these limits exempts the receiver from classification as a munition. Vendor interpretations differ. The rule refers to operation at both the target altitude and speed, but some receivers stop operating even when stationary. This has caused problems with some amateur radio balloon launches that regularly reach 30 kilometres (19 mi).

These limits only apply to units exported from (or which have components exported from) the USA – there is a growing trade in various components, including GPS units, supplied by other countries, which are expressly sold as ITAR-free.

Military

Attaching a GPS guidance kit to a dumb bomb, March 2003.

As of 2009, military applications of GPS include:

- Navigation: GPS allows soldiers to find objectives, even in the dark or in unfamiliar territory, and to coordinate troop and supply movement. In the United States armed forces, commanders use the *Commanders Digital Assistant* and lower ranks use the *Soldier Digital Assistant*.
- Target tracking: Various military weapons systems use GPS to track potential ground and air targets before flagging them as hostile. These weapon systems pass target coordinates to precision-guided munitions to allow them to engage targets accurately. Military aircraft, particularly in air-to-ground roles, use GPS to find targets (for example, gun camera video from AH-1 Cobras in Iraq show GPS co-ordinates that can be viewed with specialised software).
- Missile and projectile guidance: GPS allows accurate targeting of various military weapons including ICBMs, cruise missiles, precision-guided munitions and Artillery projectiles. Embedded GPS receivers able to withstand accelerations of 12,000 *g* or about 118 km/s^2 have been developed for use in 155 millimetres (6.1 in) howitzers.
- Search and Rescue: Downed pilots can be located faster if their position is known.
- Reconnaissance: Patrol movement can be managed more closely.
- GPS satellites carry a set of nuclear detonation detectors consisting of an optical sensor (Y-sensor), an X-ray sensor, a

dosimeter, and an electromagnetic pulse (EMP) sensor (W-sensor), that form a major portion of the United States Nuclear Detonation Detection System. General William Shelton has stated that this feature may be dropped from future satellites in order to save money.

Communication

The navigational signals transmitted by GPS satellites encode a variety of information including satellite positions, the state of the internal clocks, and the health of the network. These signals are transmitted on two separate carrier frequencies that are common to all satellites in the network. Two different encodings are used: a public encoding that enables lower resolution navigation, and an encrypted encoding used by the U.S. military.

Message Format

GPS message format	
Subframes	***Description***
1	Satellite clock, GPS time relationship
2–3	Ephemeris (precise satellite orbit)
4–5	Almanac component (satellite network synopsis, error correction)

Each GPS satellite continuously broadcasts a *navigation message* on L1 C/A and L2 P/Y frequencies at a rate of 50 bits per second. Each complete message takes 750 seconds (12 1/2 minutes) to complete. The message structure has a basic format of a 1500-bit-long frame made up of five subframes, each subframe being 300 bits (6 seconds) long. Subframes 4 and 5 are subcommutated 25 times each, so that a complete data message requires the transmission of 25 full frames. Each subframe consists of ten words, each 30 bits long. Thus, with 300 bits in a subframe times 5 subframes in a frame times 25 frames in a message, each message is 37,500 bits long. At a transmission rate of 50-bit/s, this gives 750 seconds to transmit an entire almanac message (GPS). Each 30-second frame begins precisely on the minute or half-minute as indicated by the atomic clock on each satellite.

The first subframe of each frame encodes the week number and the time within the week, as well as the data about the health of the satellite. The second and the third subframes contain the *ephemeris* –

the precise orbit for the satellite. The fourth and fifth subframes contain the *almanac*, which contains coarse orbit and status information for up to 32 satellites in the constellation as well as data related to error correction. Thus, in order to obtain an accurate satellite location from this transmitted message the receiver must demodulate the message from each satellite it includes in its solution for 18 to 30 seconds. In order to collect all the transmitted almanacs the receiver must demodulate the message for 732 to 750 seconds or 12 1/2 minutes.

All satellites broadcast at the same frequencies. Signals are encoded using code division multiple access (CDMA) allowing messages from individual satellites to be distinguished from each other based on unique encodings for each satellite (that the receiver must be aware of). Two distinct types of CDMA encodings are used: the coarse/acquisition (C/A) code, which is accessible by the general public, and the precise (P(Y)) code, which is encrypted so that only the U.S. military and other NATO nations who have been given access to the encryption code can access it.

The ephemeris is updated every 2 hours and is generally valid for 4 hours, with provisions for updates every 6 hours or longer in non-nominal conditions. The almanac is updated typically every 24 hours. Additionally, data for a few weeks following is uploaded in case of transmission updates that delay data upload.

Problem Description

The receiver uses messages received from satellites to determine the satellite positions and time sent. The *x, y,* and *z* components of satellite position and the time sent are designated as $[x_i, y_i, z_i, t_i]$ where the subscript i denotes the satellite and has the value 1, 2, ..., n, where $n \geq 4$. When the time of message reception indicated by the on-board clock is $\tilde{t}_{\mathrm{ri}}$, the true reception time is $t_{\mathrm{ri}} = \tilde{t}_{\mathrm{ri}} + b$ where b is receiver's clock offset from GPS system employed by the satellites. The receiver clock offset is the same for all received satellite signals. The message's transit time is $\tilde{t}_{\mathrm{ri}} + b - t_i$. Assuming the message travelled at the speed of light, c, the distance travelled is $\left(\tilde{t}_{\mathrm{ri}} + b - t_i\right)c$.

If the distance travelled between the receiver and satellite i and the distance travelled between the receiver and satellite j are subtracted, the result is $\left(\tilde{t}_{\mathrm{ri}} - t_i\right)c - \left(\tilde{t}_{\mathrm{rj}} - t_j\right)c$, which only involves known or measured quantities. The locus of points having a constant difference in distance to two points (here, two satellites) is a hyperboloid. Thus, from four measured reception times, the receiver can be placed at the intersection of the surfaces of three hyperboloids. When more than four satellites are being utilized, in the ideal case of no errors, the receiver is at the intersection of the surfaces of n-1 hyperboloids, where n is the number of satellites.

If the satellites were stationary, then three equations describing hyperbolas could be solved simultaneously to derive the receiver position. This would be analogous to the calculations performed for terrestrial hyperbolic systems such as LORAN-C. However, since the satellites are in motion and broadcast their locations based on GPS system time, the receiver clock offset from GPS system time must also be determined.

The clock error or bias, *b*, is the amount that the receiver's clock is offset from GPS system (*satellite*) time. In the case of four satellites, the receiver has four unknowns, the three components of GPS receiver position and the clock bias [*x, y, z, b*]. Thus the equations to be solved are:

$$(x-x_i)^2+(y-y_i)^2+(z-z_i)^2=\left([\tilde{t}_{\mathrm{r}}+b-t_i]c\right)^2,\ i=1,2,\ldots,n$$

or in terms of *pseudoranges*, $p_i=\left(\tilde{t}_{\mathrm{r}}-t_i\right)c$, as

$$p_i=\sqrt{(x-x_i)^2+(y-y_i)^2+(z-z_i)^2}-bc,\ i=1,2,\ldots,n.$$

These equations can be solved by algebraic or numerical methods.

Least Squares Solution Method

When four or more satellites are available, the calculation can use the four best, or more than four (up to all visible satellites), considering number of receiver channels, processing capability, and geometric dilution of precision (GDOP). Using more than four involves an over-determined system of equations with no unique solution; such a system can be solved by a least-squares or weighted least squares method (e.g., Gauss–Newton algorithm). Errors can be estimated through the residuals. With each combination of satellites, GDOP quantities can be calculated based on the relative sky directions of the satellites used. The receiver location is expressed in a specific coordinate system, such as latitude and longitude using the WGS 84 geodetic datum or a country-specific system.

$$\left(\hat{x},\hat{y},\hat{z},\hat{b}\right)=\underset{(x,y,z,b)}{\arg\min}\sum_i\left(\sqrt{(x-x_i)^2+(y-y_i)^2+(z-z_i)^2}-bc-p_i\right)^2$$

The GPS system was initially developed assuming use of a least-squares solution method — i.e., before closed-form solutions were found.

Closed-form Solution Methods (Bancroft, etc.)

The first closed-form solution to the above set of equations was discovered by S. Bancroft. Its properties are well known. Bancroft's method is algebraic, as opposed to numerical, and can be used for four or more satellites. When four satellites are used, the key steps are inversion of a 4x4 matrix and solution of a single-variable quadratic equation. Bancroft's method provides one or two solutions for the unknown quantities. When there are two (usually the case), only one will be a near-earth sensible solution.

When more than four satellites are to be used for a solution, then Bancroft uses the Generalized inverse (i.e., the pseudoinverse) to find a solution. However, a case has been made that iterative methods (e.g., Gauss-Newton algorithm) for solving over-determined Non-linear least squares(NLLS) problems generally provide more accurate solutions.

Other closed-form solutions were published after Bancroft. Their use in practice is unclear.

Error Sources and Analysis

GPS error analysis examines the sources of errors in GPS results and the expected size of those errors. GPS makes corrections for receiver clock errors and other effects but there are still residual errors which are not corrected. Sources of error include signal arrival time measurements, numerical calculations, atmospheric effects (ionospheric/tropospheric delays), ephemeris and clock data, multipath signals, and natural and artificial interference. The magnitude of the residual errors resulting from these sources is dependent on geometric dilution of precision. Artificial errors may result from jamming devices and threaten ships and aircraft or from intentional signal degradation through selective availability, which limited accuracy to ~6–12 m, but which has now been switched off.

Accuracy Enhancement and Surveying

Augmentation

Integrating external information into the calculation process can materially improve accuracy. Such augmentation systems are generally named or described based on how the information arrives. Some systems transmit additional error information (such as clock drift, ephemera, or ionospheric delay), others characterize prior errors, while a third group provides additional navigational or vehicle information.

Examples of augmentation systems include the Wide Area Augmentation System (WAAS), European Geostationary Navigation

Overlay Service (EGNOS), Differential GPS (DGPS), Inertial Navigation Systems (INS) and Assisted GPS. The standard accuracy of about 15 metres (49 feet) can be augmented to 3–5 metres (9.8–16.4 ft) with DGPS, and to about 3 metres (9.8 feet) with WAAS.

Precise Monitoring

Accuracy can be improved through precise monitoring and measurement of existing GPS signals in additional or alternate ways.

The largest remaining error is usually the unpredictable delay through the ionosphere. The spacecraft broadcast ionospheric model parameters, but some errors remain.

This is one reason GPS spacecraft transmit on at least two frequencies, L1 and L2. Ionospheric delay is a well-defined function of frequency and the total electron content (TEC) along the path, so measuring the arrival time difference between the frequencies determines TEC and thus the precise ionospheric delay at each frequency.

Military receivers can decode the P(Y) code transmitted on both L1 and L2. Without decryption keys, it is still possible to use a *codeless* technique to compare the P(Y) codes on L1 and L2 to gain much of the same error information. However, this technique is slow, so it is currently available only on specialised surveying equipment. In the future, additional civilian codes are expected to be transmitted on the L2 and L5 frequencies. Then all users will be able to perform dual-frequency measurements and directly compute ionospheric delay errors.

A second form of precise monitoring is called *Carrier-Phase Enhancement* (CPGPS). This corrects the error that arises because the pulse transition of the PRN is not instantaneous, and thus the correlation (satellite-receiver sequence matching) operation is imperfect. CPGPS uses the L1 carrier wave, which has a period of $\frac{1\text{s}}{1575.42\times10^6} = 0.63475\text{ns} \approx 1\text{ns}$, which is about one-thousandth of the C/A Gold code bit period of $\frac{1\text{s}}{1023\times10^3} = 977.5\text{ns} \approx 1000\text{ns}$, to act as an additional clock signal and resolve the uncertainty. The phase difference error in the normal GPS amounts to 2–3 metres (6.6–9.8 ft) of ambiguity. CPGPS working to within 1% of perfect transition reduces this error to 3 centimetres (1.2 in) of ambiguity. By eliminating this error source, CPGPS coupled with DGPS normally realises between 20–30 centimetres (7.9–11.8 in) of absolute accuracy.

Relative Kinematic Positioning (RKP) is a third alternative for a precise GPS-based positioning system. In this approach, determination of range signal can be resolved to a precision of less than 10 centimetres (3.9 in).

This is done by resolving the number of cycles that the signal is transmitted and received by the receiver by using a combination of differential GPS (DGPS) correction data, transmitting GPS signal phase information and ambiguity resolution techniques via statistical tests—possibly with processing in real-time (real-time kinematic positioning, RTK).

Timekeeping

Leap Seconds: While most clocks derive their time from Coordinated Universal Time (UTC), the atomic clocks on the satellites are set to GPS time (GPST). The difference is that GPS time is not corrected to match the rotation of the Earth, so it does not contain leap seconds or other corrections that are periodically added to UTC. GPS time was set to match UTC in 1980, but has since diverged. The lack of corrections means that GPS time remains at a constant offset with International Atomic Time (TAI) (TAI " GPS = 19 seconds). Periodic corrections are performed to the on-board clocks to keep them synchronized with ground clocks.

The GPS navigation message includes the difference between GPS time and UTC. As of July 2012, GPS time is 16 seconds ahead of UTC because of the leap second added to UTC June 30, 2012. Receivers subtract this offset from GPS time to calculate UTC and specific timezone values. New GPS units may not show the correct UTC time until after receiving the UTC offset message. The GPS-UTC offset field can accommodate 255 leap seconds (eight bits).

Accuracy

GPS time is theoretically accurate to about 14 nanoseconds. However, most receivers lose accuracy in the interpretation of the signals and are only accurate to 100 nanoseconds.

Format

As opposed to the year, month, and day format of the Gregorian calendar, the GPS date is expressed as a week number and a seconds-into-week number. The week number is transmitted as a ten-bit field in the C/A and P(Y) navigation messages, and so it becomes zero again every 1,024 weeks (19.6 years). GPS week zero started at 00:00:00 UTC

(00:00:19 TAI) on January 6, 1980, and the week number became zero again for the first time at 23:59:47 UTC on August 21, 1999 (00:00:19 TAI on August 22, 1999).

To determine the current Gregorian date, a GPS receiver must be provided with the approximate date (to within 3,584 days) to correctly translate the GPS date signal. To address this concern the modernised GPS navigation message uses a 13-bit field that only repeats every 8,192 weeks (157 years), thus lasting until the year 2137 (157 years after GPS week zero).

Carrier Phase Tracking (Surveying)

Another method that is used in surveying applications is carrier phase tracking. The period of the carrier frequency multiplied by the speed of light gives the wavelength, which is about 0.19 metres for the L1 carrier. Accuracy within 1% of wavelength in detecting the leading edge reduces this component of pseudorange error to as little as 2 millimetres. This compares to 3 metres for the C/A code and 0.3 metres for the P code.

However, 2 millimetre accuracy requires measuring the total phase—the number of waves multiplied by the wavelength plus the fractional wavelength, which requires specially equipped receivers. This method has many surveying applications. It is accurate enough for real-time tracking of the very slow motions of tectonic plates, typically 0–100 mm (0–4 inches) per year.

Triple differencing followed by numerical root finding, and a mathematical technique called least squares can estimate the position of one receiver given the position of another. First, compute the difference between satellites, then between receivers, and finally between epochs. Other orders of taking differences are equally valid. Detailed discussion of the errors is omitted.

The satellite carrier total phase can be measured with ambiguity as to the number of cycles. Let $\phi(r_i, s_j, t_k)$ denote the phase of the carrier of satellite *j* measured by receiver *i* at time t_k. This notation shows the meaning of the subscripts *i, j,* and *k*. The receiver (*r*), satellite (*s*), and time (*t*) come in alphabetical order as arguments of ϕ and to balance readability and conciseness, let $\phi_{i,j,k} = \phi(r_i, s_j, t_k)$ be a concise abbreviation. Also we define three functions, : $\Delta^r, \Delta^s, \Delta^t$, which return differences between receivers, satellites, and time points, respectively. Each function has variables with three subscripts as its arguments.

These three functions are defined below. If $\alpha_{i,j,k}$ is a function of the three integer arguments, *i, j,* and *k* then it is a valid argument for the functions, $\Delta^r, \Delta^s, \Delta^t$:, with the values defined as

$$\Delta^r(\alpha_{i,j,k}) = \alpha_{i+1,j,k} - \alpha_{i,j,k},$$

$$\Delta^s(\alpha_{i,j,k}) = \alpha_{i,j+1,k} - \alpha_{i,j,k}, \text{ and}$$

$$\Delta^t(\alpha_{i,j,k}) = \alpha_{i,j,k+1} - \alpha_{i,j,k} .$$

Also if $\alpha_{i,j,k}$ *and* $\beta_{l,m,n}$ are valid arguments for the three functions and *a* and *b* are constants then $(a\,\alpha_{i,j,k} + b\,\beta_{l,m,n})$ is a valid argument with values defined as

$$\Delta^r(a\,\alpha_{i,j,k} + b\,\beta_{l,m,n}) = a\,\Delta^r(\alpha_{i,j,k}) + b\,\Delta^r(\beta_{l,m,n}),$$

$$\Delta^s(a\,\alpha_{i,j,k} + b\,\beta_{l,m,n}) = a\,\Delta^s(\alpha_{i,j,k}) + b\,\Delta^s(\beta_{l,m,n}), \text{ and}$$

$$\Delta^t(a\,\alpha_{i,j,k} + b\,\beta_{l,m,n}) = a\,\Delta^t(\alpha_{i,j,k}) + b\,\Delta^t(\beta_{l,m,n}) .$$

Receiver clock errors can be approximately eliminated by differencing the phases measured from satellite 1 with that from satellite 2 at the same epoch. This difference is designated as

Double differencing computes the difference of receiver 1's satellite difference from that of receiver 2. This approximately eliminates satellite clock errors. This double difference is:

$$\Delta^s(\phi_{1,1,1}) = \phi_{1,2,1} - \phi_{1,1,1}$$

Triple differencing subtracts the receiver difference from time 1 from that of time 2. This eliminates the ambiguity associated with the integral number of wavelengths in carrier phase provided this ambiguity does not change with time. Thus the triple difference result eliminates practically all clock bias errors and the integer ambiguity. Atmospheric delay and satellite ephemeris errors have been significantly reduced. This triple difference is:

$$\Delta^r(\Delta^s(\phi_{1,1,1})) \quad =\Delta^r(\phi_{1,2,1} - \phi_{1,1,1}) \quad =\Delta^r(\phi_{1,2,1}) - \Delta^r(\phi_{1,1,1}) \quad =(\phi_{2,2,1} - \phi_{1,2,1}) - (\phi_{2,1,1} - \phi_{1,1,1})$$

Triple difference results can be used to estimate unknown variables. For example if the position of receiver 1 is known but the position of receiver 2 unknown, it may be possible to estimate the position of receiver 2 using numerical root finding and least squares.

$$\Delta^t(\Delta^r(\Delta^s(\phi_{1,1,1})))$$

Triple difference results for three independent time pairs quite possibly will be sufficient to solve for receiver 2's three position components. This may require the use of a numerical procedure. An approximation of receiver 2's position is required to use such a numerical method.

This initial value can probably be provided from the navigation message and the intersection of sphere surfaces. Such a reasonable estimate can be key to successful multidimensional root finding. Iterating from three time pairs and a fairly good initial value produces one observed triple difference result for receiver 2's position. Processing additional time pairs can improve accuracy, overdetermining the answer with multiple solutions. Least squares can estimate an overdetermined system. Least squares determines the position of receiver 2 which best fits the observed triple difference results for receiver 2 positions under the criterion of minimizing the sum of the squares.

Bibliography

Berry, J.K.: *Beyond Mapping: Concepts, Algorithms and Issues in GIS*, GIS World Books, Fort Collins, CO, 1993.

Bolstad, P.: *GIS Fundamentals: A first text on Geographic Information Systems*, Eider Press, White Bear Lake, MN, 2005.

Bomford, Brigadier G.: *Geodesy*; Clarendon Press; London, England, 1 952.

Brown, D.C.: *Doppler Positioning by the Short Arc Method*; DBA Systems, Inc.; Melbourne, Florida; 1976.

Brown, L.A.: *A Story of Maps*; Little, Brown and Company; Boston, Massachusetts; 1950.

Burrough, P.A. and McDonnell, R.A.: *Principles of Geographical Information Systems*, Oxford University Press, Oxford, 1998.

Chang, K.: *Introduction to Geographic Information System*, McGraw Hill, New York, 2007.

Clark, D.: *Plane and Geodetic Surveying*; Constable and Company, Ltd.; London, England; 1957.

de Smith M J, Goodchild M F, Longley P.A.: *Geospatial Analysis: A Comprehensive Guide to Principles, Techniques and Software Tools*, Troubador, UK. 2007.

Elangovan, K.: *GIS: Fundamentals, Applications and Implementations*, New India Publishing Agency, New Delhi, 2006.

Ewing, C.E. and M.M. Mitchell: *Introduction to Geodesy*; American Elsevier Publishing Company, Inc.; New York, 1970.

Fu, P., and J. Sun: *Web GIS: Principles and Applications*. ESRI Press. Redlands, CA. 2010..

Gaposchkin, E.M.: *Smithsonian Standard Earth (III); Smithsonian Institution Astrophysical Observatory*; Cambridge, Massachusetts; 1973.

Gross, J.E.: *Preprocessing Electronic Satellite Observations*; The Ohio State University; Columbus, Ohio; 1968.

Guier, W.H. and G.C. Weiffenbach: *"A Satellite Doppler Navigational System"*; Proceedings of the IRE; , 1960.

Harvey, Francis: *A Primer of GIS, Fundamental Geographic and Cartographic Concepts.* The Guilford Press, 2008.

Heiskanen, W.A. and F.A.: *Vening Meinesz; The Earth and Its Gravity Field*; McGraw-Hill Book Company, Inc.; New York, 1958.

Heywood, I., Cornelius, S., and Carver, S.: *An Introduction to Geographical Information Systems*, Prentice Hall, NY, 2006.

Hosmer, G.L.: *Geodesy*; John Wiley and Sons, Inc.; New York, 1956.

Kaula, W. M.: *Theory of Satellite Geodesy*; Blaisdell Publishing Company; Waltham, Massachusetts; 1966.

Longley, P.A., Goodchild, M.F., Maguire, D.J. and Rhind, D.W.: *Geographic Information Systems and Science*, Wiley, Chichester, 2005.

Maguire, D.J., Goodchild M.F., Rhind D.W.: *"Geographic Information Systems: principles, and applications"*, Longman Scientific and Technical, Harlow, 1997,

Meade, B.K.: *"High-Precision Transcontinental Traverse Surveys in the United States"*; National Oceanic and Atmospheric Administration; Rockville, Maryland; 1969.

Mueller, I.I. and J.D. Rockie: *Gravimetric and Celestial Geodesy*; Frederick Ungar Publishing Co., New York; 1966.

Ott, T. and Swiaczny, F.: *Time-integrative GIS. Management and Analysis of Spatio-temporal Data*, Springer, Berlin / Heidelberg / New York, 2001.

Phillips, J.O.: *Specifications to Support Classification, Standards of Accuracy, and General Specifications of Geodetic Control Surveys*; Federal Geodetic Control Committee; NationalOceanic and Atmospheric Administration; Rockville, Maryland; 1980.

Schmidt, Captain J.R.: *"Computer Error Analysis of Tropospheric Effects for the NAVSTAR Global Positioning System"*; Air Force Institute of Technology; Wright-Patterson AFB, Ohio; 1975.

Shapiro, I.I., and C.F. Martin: *"Precision Geodesy Via Radio Interferometry: First Results";* Massachusetts Institute of Technology; Cambridge, Massachusetts; 1971.

Thurston, J., Poiker, T.K. and J. Patrick Moore: *Integrated Geospatial Technologies: A Guide to GPS, GIS, and Data Logging*, Wiley, Hoboken, New Jersey, 2003.

Wells, D.E.: *"Semi Dynamical Doppler Satellite Positioning"*; Bedford Institute of Oceanography; Dartmouth, Nova Scotia, Canada; August 1975.

Wheatley, David and Gillings, Mark: *Spatial Technology and Archaeology. The Archaeological Application of GIS*, Taylor & Francis, London, New York, 2002.

Whitmore, G.D.: *Advanced Surveying and Mapping*; International Textbook Company; Scranton, Pennsylvania; 1952.

Wise, S.: *GIS Basics*, Taylor & Francis, London, 2002.

Worboys, Michael; Duckham, Matt: *GIS: a Computing perspective*, CRC Press, Boca Raton, 2004.

Index

❑❑❑